Principles and Applications of
HOMOGENEOUS CATALYSIS

Akira Nakamura, Professor

Department of Polymer Science
Faculty of Science
Osaka University
Osaka, Japan

Minoru Tsutsui

Professor of Chemistry
Department of Chemistry
Texas A&M University
College Station, Texas 77843 U.S.A.

A WILEY-INTERSCIENCE PUBLICATION

JOHN WILEY & SONS
New York   Chichester   Brisbane   Toronto

Copyright © 1980 by John Wiley & Sons, Inc.

All rights reserved. Published simultaneously in Canada.

Reproduction or translation of any part of this work beyond that permitted by Sections 107 or 108 of the 1976 United States Copyright Act without the permission of the copyright owner is unlawful. Requests for permission or further information should be addressed to the Permissions Department, John Wiley & Sons, Inc.

Library of Congress Cataloging in Publication Data:

Nakamura, Akira, 1934–
   Principles and applications of homogeneous catalysis.

   "A Wiley–Interscience publication."
   1. Catalysis. I. Tsutsui, Minoru, 1918– joint author. II. Title.

QD505.N36   541'.395   79-24754
ISBN 0-471-02869-X

Printed in the United States of America

10 9 8 7 6 5 4 3 2 1

PREFACE

Homogeneous catalysis is becoming an indispensable discipline in chemistry. A practical knowledge of homogeneous catalysis and an understanding of its basic principles are necessary for senior researchers and for graduate students. Since catalysis in many industrially important new processes utilizes transition metal species as active catalysts in homogeneous solutions, the study of metal complexes has constituted the mainstream of research in homogeneous catalysis. Therefore this book focuses on the chemistry of metal complexes pertaining to catalysis. We have tried to systematize this rapidly expanding field and have included relevant acid-base catalysts, organic catalysts, and enzymes.

It is assumed that the reader has knowledge of organic, inorganic, and physical chemistry at a graduate level. Since the inorganic chemistry of metal complexes has developed separately from general organic and biological chemistry, we feel that meaningful correlations of these major chemical sciences might be made by a systematic and comprehensive study of homogeneous catalysis.

Heterogeneous catalysis utilizing reactive surfaces of transition metals or metal oxides is important in many chemical reactions. The utility of heterogeneous catalysis ranges from large-scale cracking of oil to laboratory-scale hydrogenations. Often, however, it is not highly selective, and it sometimes requires high temperature and pressures, causing a waste of starting materials and energy. Any waste is an economic penalty in an age of increasing costs of raw materials and energy. Because of the complexity of surface phenomena, rational improvement of the efficiency of heterogeneous catalysts remains a formidable problem.

Although homogeneous catalysts have been successful in catalyzing about 20 major industrial processes as well as numerous small-scale reactions with high selectivity, there are still two major obstacles to overcome. One is catalyst separation following reactions. The other is the application of homogeneous catalysts to large-scale, technologically important reactions such as coal liquefaction, the water-gas shift reaction, and ammonia synthesis, where heterogeneous catalysis prevails. The separation problem is now being approached by immobilizing the catalytic metal complex molecules on synthetic high polymers or inorganic catalyst supports. These large-scale, technologically important reactions are now also a target of homogeneous catalysts to effect such industrial reactions by combining three or more metal atoms in clusters where such reactions have been found to take place at least stoichiometrically. At the present stage, only a few transition metal clusters, such as $Ru_3(CO)_{12}$ or $Ir_4(CO)_{12}$, are known to catalyze on a laboratory scale the water-gas shift reaction ($H_2O + CO \rightarrow H_2 + CO_2$) or the methanation of carbon monoxide. Since rational modifications of metal complexes or metal cluster compounds will become feasible in the future, homogeneous catalysis may take over many important reactions with its characteristic features of high selectivity, activity, and versatility.

This book would never have been completed without the assistance, useful discussions, and proofreading of Professor J. J. Lagowski, Professor D. E. Bergbreiter, Dr. J. Francis, Dr. J. Unruh and our colleagues Dr. Paul Roling, Dr. Darrell Axtell, Dr. Rex Bobsein, Dr. K. Tatsumi, Miss Arlene Courtney, Dr. Duane Hrncir, and others. And thanks to Ruth Lathem for her typing of the manuscript. The time expended in writing this book was supported partly by the Japan Society of the Promotion of Science.

AKIRA NAKAMURA
MINORU TSUTSUI

Osaka, Japan
College Station, Texas
December 1979

Contents

1. **Importance**

    1. Present Status and Outlook, 1
    2. Academic and Industrial Problems, 6

2. **Characteristic Features**

    1. Homogeneity, 9
    2. Specificity and Selectivity, 11

3. **Basic Principles**

    1. Chemical Kinetics and Energetic Aspects, 14
        - 1.1. Zeroth-Order Reactions, 14
        - 1.2. First-Order Reactions, 15
        - 1.3. Pseudo-First-Order Reactions, 16
        - 1.4. Complex Reactions, 16
        - 1.5. The Michaelis-Menten Rate Equation, 18
        - 1.6. Rate Dependence on Catalyst Concentration, 20
        - 1.7. The Rate-Determining Step, 20
        - 1.8. Activation Parameters, 22
    2. Homogeneous Active Sites: Activation and Deactivation, 24
    3. Fundamental Aspects of Selectivity, 25
    4. Stereochemistry, Orbital Symmetry, and Reactivity, 32

4. **Elementary Processes**

    1. General Interactions, 45
        - 1.1. Electrophilic and Nucleophilic Interactions, 45

1.2. Electron Donor-Acceptor Interactions, 52
1.3. Radical Interactions, 56

2. Elementary Reactions in Transition Metal Chemistry, 60
   2.1. Coordination and Dissociation of Ligands, 60
   2.2. Oxidative Addition and Reductive Elimination, 64
   2.3. Insertion and De-insertion, 75
   2.4. Cycloadditions and Electrocyclic Reactions, 81
   2.5. Fluxionality and Polytopal Rearrangements, 88
   2.6. Reactions of Coordinated Ligands, 92
   2.7. σ-π Rearrangements, 97

5. Mechanisms

1. Hydrolysis and Condensation, 109
2. Polar Addition and Ionic Elimination, 115
3. Electron Transfer Redox Reactions, 119
4. Group Transfer Reactions, 122
5. Methyl Transfer Reactions, 122
6. Isomerizations, 123
7. Hydrogenations, 125
8. Carbonylations, 130
9. Oligomerizations, 133
10. Polymerizations, 139
11. Oxidations, 145
12. Carbenoid Reactions, 150
13. Nitrogen Fixations, 154
14. Enantioselective Catalytic Reactions, 159

6. Further Developments

1. Heterogenized Homogeneous Catalysts, 173
2. Enzyme-like Catalysts, 176
3. Extremely Selective Catalysts, 179

Contents

    4. Energy Problems, 181

    5. Phase-Transfer Catalysts, 182

7. Industrial Applications

    1. Petrochemicals, 187

    2. Coal, 190

    3. Fine Chemicals, 192

Index 197

Principles and Applications of
HOMOGENEOUS CATALYSIS

1

IMPORTANCE

1. PRESENT STATUS AND OUTLOOK

The field of homogeneous catalysis includes vast areas of active research ranging from simple acid-base catalysis to extremely complex metalloenzyme catalysis. It is one of the most rapidly expanding fields of chemistry: the recent development of homogeneous catalysis has been so fast that its definition and scope remain to be settled.

Before 1960 only a few homogeneous catalysts were used, intermittently, on an academic or an industrial scale. The past several years, however, have seen the emergence of a variety of novel, useful, and important homogeneous catalyst systems, and this rapid enrichment seems to be only the beginning of further growth of this field.

Let us take a closer look at some of these developments. Specific acids (H^+) and bases (OH^-) are the simplest homogeneous catalysts,* whose activities are limited to special substrates because no modification of the catalytic species is possible. Classical studies in physical chemistry have revealed the phenomena of both general acid (electrophilic) and base (nucleophilic) catalysis.† Here, structural modifications are possible, and a high activity for polar reactions has been recently realized, in particular with multifunctional acid-base catalysts.

---

* *Specific acid catalysts* involve H_3O^+ as the sole catalytically effective species.
† In *general acid catalysis,* all the protic acids (Brønsted acids) present in the reaction mixture participate in the catalysis.

Relevant metal ions have been found to function as superacids (or superelectrophiles) to accelerate some reactions to a great extent. Redox properties of some transition metal ions (e.g., Cu^+ and Cu^{2+}) have been advantageously utilized to catalyze electron transfer reactions. The development of coordination chemistry has greatly helped the understanding of these metal ion catalysts and has intensified the search for catalytically active metal systems exploiting novel and unique ligands.

Enzymes also belong to the family of homogeneous catalysts. The development of enzyme research has followed a different course because of its inherent biochemical characteristics. Enzyme action was once regarded as mystical, transcending known laws of chemistry and physics. The elucidation of the catalytic action of some enzymes by ultramodern scientific techniques such as x-ray structural analysis, nuclear magnetic resonance (nmr), and electron paramagnetic resonance (epr) has clearly revealed that enzymes are homogeneous catalysts whose action can be understood by established principles of chemistry.

The rapid development of organometallic chemistry is enormously indebted to the organometallic catalysts used in three industrially important reactions: the Ziegler, Wacker, and "oxo" processes.

1. The Ziegler catalysis:

$$CH_2 = CH_2 \xrightarrow[\text{1 atm, room temp.}]{TiCl_4/AlEt_3} -CH_2-CH_2-CH_2-CH_2-$$

2. The Wacker process:

$$CH_2 = CH_2 + O_2 \xrightarrow[120-130°]{PdCl_2/CuCl_2/aq\ HCl} CH_3CHO$$

3. The "oxo" process:

$$CH_2 = CH_2 + CO + H_2 \xrightarrow[140°,\ 100\ atm]{Co_2(CO)_8} CH_3CH_2CHO$$

Delicate combinations of ligand steric and electronic effects have been found to influence strongly the structure and reactivity of labile, catalytically active organometallic complexes. With an increasing understanding of these organometallic compounds, attempts have been made to construct the exact structure that is needed for a particular reaction. As this structural tailoring process of metal complexes proceeds, the active metal complexes resemble metalloenzymes more and more closely (Fig. 1a). In this light, metalloenzymes can be regarded

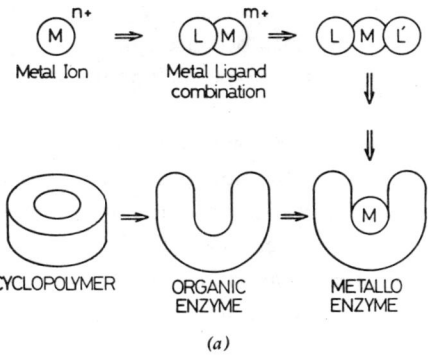

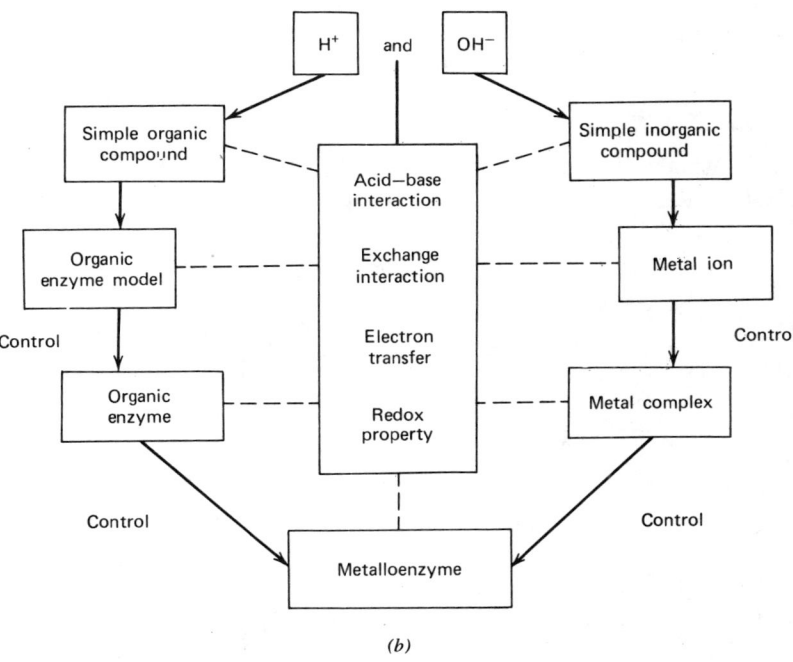

Fig. 1. (*a*) The development of homogeneous catalysts: an idealized scheme. (*b*) The development and correction of homogeneous catalysts.

as naturally occurring tailor-made metal complexes. The same point of view holds for the nonmetallic enzymes because synthetic organic catalysts (e.g., cyclodextrins or micelles) are simply mimicking the structure and function of enzyme active sites.

Although the interactions between substrates and catalysts are more complex in enzymes or similar catalysts, the fundamentally important interactions are simple: (*a*) acid-base (or more generally donor-acceptor) interactions, (*b*) electronic exchange (including weak van der Waals attraction) interactions, and (*c*) electron transfer (electron donor-acceptor) interactions. Using these three types of interaction, various homogeneous catalysts can be correlated according to Fig. 1*b*.

Starting from the simplest and totally symmetric homogeneous catalysts H^+ and OH^-, steric and electronic control in catalysis has been promoted by utilizing surface holes, pits, crevices, or particular metal ions or atoms. The most specific control of reactions may be found in the most complex homogeneous catalysts, namely, enzymes. However recent progress in the investigation of models of organic enzymes or metalloenzymes has developed materials that are as selective as natural enzymes.

We expect the establishment of firmer correlations among family members of homogeneous catalysts in the future. Paths for the development of particular homogeneous catalysts are described in Chapter 6.

Fundamental differences among catalytically active metal complexes, organic catalysts, and organic enzymes are illustrated in Figs. 2 through 4.

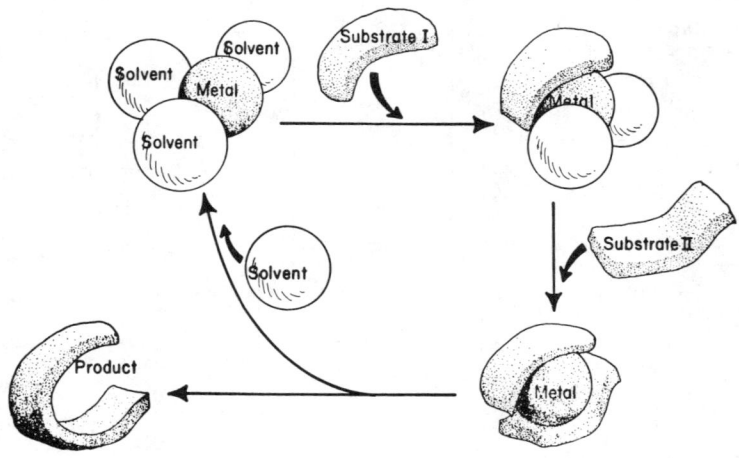

Fig. 2. A schematic drawing of metal-atom catalysis.

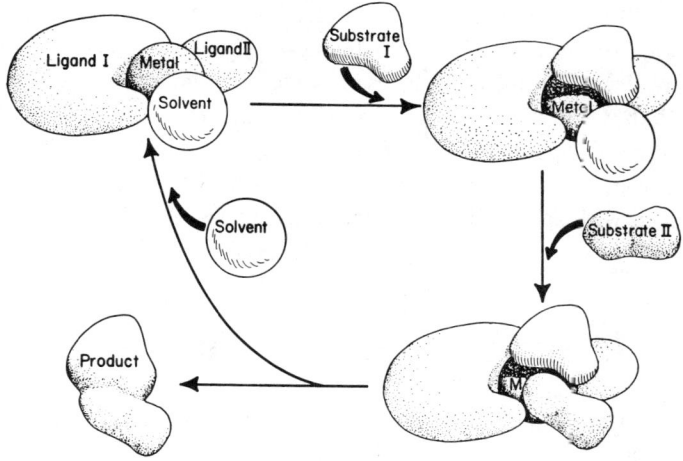

Fig. 3. A schematic drawing of metal-ligand catalysis.

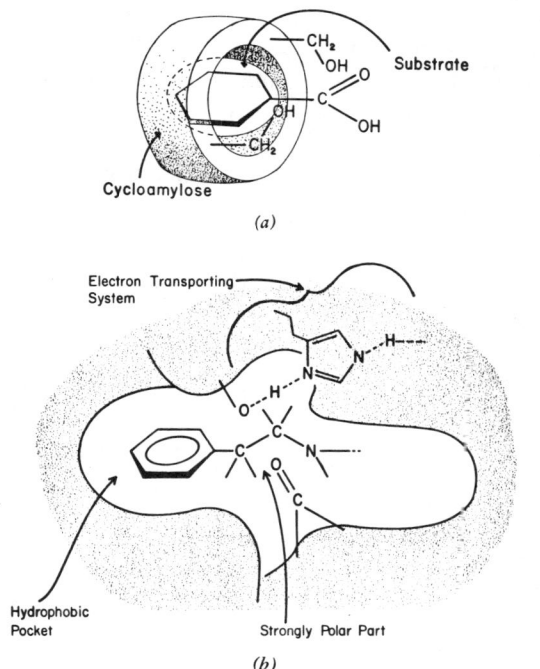

Fig. 4. (a) Cycloamylose as an enzyme model. (b) A schematic drawing of a typical active site of an enzyme.

As these figures indicate, metal complex catalysts work by arranging substrates in appropriate order around the metal active site. In contrast, organic catalysts fit the substrate into its holes or crevices and induce the desired reaction (Fig. 4). Both types of catalyst accomplish the same result using different means. By combining the ligand ordering of metal complex catalysts with the steric and hydrophobic or hydrophilic topologies of enzymes, catalysts conceivably could be produced with selectivities and activities superior to those of natural enzymes.

2. ACADEMIC AND INDUSTRIAL PROBLEMS

Although well-known homogeneous catalytic systems such as the Wacker or oxo processes have been successfully used on an industrial scale, there are ever increasing problems associated with these catalytic reactions. Because the chemical market changes frequently for many unforeseeable reasons, the importance of raw materials fluctuates. Today, a clean chemical reaction is a "must," not only for industrial chemists but also for laboratory investigators. Saving energy and material can be accomplished by utilizing new highly active, selective catalysts. To meet these requirements, homogeneous catalysis has already been given sufficient credit. For example, selective oligomerization of butadiene or allene has afforded hitherto unavailable but useful organic chemicals in large amounts (see Scheme 1 and Chapter 5).

The use of a palladium complex has replaced the mercury ion as an oxidative catalyst. Homogeneous catalysts have become familiar in the organic laboratory. In academic laboratories, because of its activity and convenience, $RhCl(PPh_3)_3$ is now the catalyst of choice for the hydrogenation of olefins or acetylenes at normal pressure and temperature (see Chapter 5). Epoxides are usually prepared

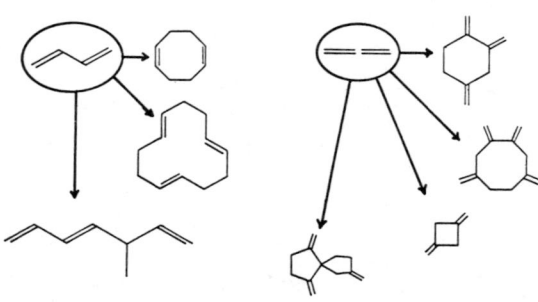

Scheme 1

for an olefin and H_2O_2 by soluble molybdenum or tungsten complexes [e.g., $MoO_2(acac)_2$ or $(MoO(O_2)_2HMPA)$] as catalysts.

$$C_2H_4 \xrightarrow[O_2/aq\ HCl]{PdCl_2/CuCl_2} CH_3CHO \xleftarrow{Hg^{2+}/H_2O} C_2H_2$$

$$R-CH=CHR' \xrightarrow[MoO(O_2)_2HMPA]{H_2O_2} R(H)C\underset{O}{-}C(H)R'$$

in hexamethylphosphoramide (HMPA)

However not all homogeneous catalysts are superior in every respect. For example, the production of only one enantiomer catalyzed by chiral metal complexes is in its formative stage (see below), and further improvement is necessary before it can be utilized as a general method.

$$PhCH=CH_2 + N_2CHCO_2Et \xrightarrow[\text{chiral copper(II) or cobalt(II) complex}]{0°} \left\{ \begin{array}{c} \text{Ph} \triangle CO_2R \\ \text{Ph} \triangle CO_2R \end{array} \right.$$

Maximum optical yield, 88%; average chemical yield, 90%

Thus, in short, utilizing homogeneous catalysis provides more convenient and efficient ways to prepare complex organic structures in the laboratory and in the fine chemicals industry because selectivity provides economy in chemicals, energy, and manpower.

As a natural consequence of an active inquiry into the mechanism of highly selective catalysts, we hope the principles leading to more selective reactions will be revealed. Academic interest will provide the stimulus for rationalizing selectivity in terms of the established theory of steric and electronic effects. Intentional modifications of the catalysts will be executed to test these hypotheses and theories. Strategic design of selective homogeneous catalysts will follow the analysis of these modifications. The results of this research will be a more sophisticated general theory concerning the interaction between small molecules and metal complexes or enzymes.

SELECTED READINGS

General

Homogeneous Catalysis, American Chemical Society *Advances in Chemistry* Series, Vol. 70, ACS, Washington, D.C., 1968.

Homogeneous Catalysis—II, D. Forster and J. F. Roth, Eds., American Chemical Society *Advances in Chemistry* Series, Vol. 132, ACS, Washington, D.C., 1974.

M. L. Bender and L. J. Brubacher, *Catalysis and Enzyme Action,* McGraw-Hill, New York, 1973.

Catalysis: Progress in Research, F. Basolo and R. L. Burwell, Jr., Eds., Plenum, New York, 1975.

Fundamental Research in Homogeneous Catalysis, M. Tsutsui and R. Ugo, Eds., Plenum, New York, 1977.

A. E. Martell and Taqui Kham, *Homogeneous Catalysis,* Vols. I and II, Academic Press, New York, 1973.

W. P. Jencks, *Catalysis in Chemistry and Enzymology,* McGraw-Hill, New York, 1969.

Aspects of Homogeneous Catalysis, Vols. 1–3, R. Ugo, Ed., Manfredi, Milan, 1973–1976.

J. M. Davidson, in *MTP International Review of Science, Inorganic Chemistry* Series 1, Vol. 6, Part 2, M. J. Mays, Ed., Butterworths, London, 1972.

A. J. Deeming, in *MTP International Review of Science, Inorganic Chemistry,* Series 2, Vol. 9, M. L. Tobe, Ed., Butterworths, London, 1974, p. 271.

R. F. Heck, *Organotransition Metal Chemistry,* Academic Press, New York, 1974.

P. W. Jolly and G. Wilke, *Organic Chemistry of Nickel,* Vols. 1 and 2, Academic Press, New York, 1974 and 1975.

O. Aguilo, in *Organometallic Reactions,* Vol. 3, E. I. Becker and M. Tsutsui, Eds., Wiley-Interscience, New York, 1972, p. 1.

S. Carra and R. Ugo, *Inorg. Chim. Acta Rev., 1,* 1 (1967).

Transition Metals in Homogeneous Catalysis, G. N. Schrauzer, Ed., Dekker, New York, 1971.

J. K. Kochi, *Organometallic Mechanisms and Catalysis,* Academic Press, New York, 1978.

Organic Synthesis via Metal Carbonyls, I. Wender and P. Pino, Eds., Vol. 1, 1968; Vol. 2, 1977, Academic Press, New York.

B. L. Shaw and N. I. Tucker, *Organo-Transition Metal Compounds and Related Aspects of Homogeneous Catalysis,* Pergamon Texts in Inorganic Chemistry, Vol. 23, Pergamon Press, Oxford, 1973.

Coordination and Catalysis, G. Henrici-Olivé and S. Olivé, Verlag-Chemie, Weinheim, 1977.

Metalloenzymes

R. J. P. Williams, *Inorg. Chim. Acta Rev., 5,* 137 (1971).

Biological Hydroxylation Mechanisms, G. S. Boyd and R. M. S. Smellie, Eds., Academic Press, London, 1972.

Metal Ions in Biological Systems, Vol. 4, H. S. Siegel, Ed., Dekker, New York, 1977.

Biological Aspects of Inorganic Chemistry, A. W. Addison et al., Eds., Wiley, New York, 1977.

G. McLendon and A. E. Martell, *Coord. Chem. Rev., 19,* 1 (1976).

H. B. Dunford and J. S. Stillman, *Coord. Chem. Rev., 19,* 187 (1976).

2

CHARACTERISTIC FEATURES

1. HOMOGENEITY

Homogeneous catalysis derives its name from its most conspicuous feature: that is, the catalyst is in a single homogeneous phase (virtually always a liquid) as a chemical compound. This characteristic clearly differentiates it from *heterogeneous* catalysis, which usually implies solid catalysts for vapor or liquid phase reactions. Thus these two catalyst systems are apparently quite different in appearance, in experimental techniques, in theory, and in practical industrial applications. However a closer look at the basic chemistry involved in the elementary steps of these reactions serves to reveal some similarities. For example, some types of "chemisorption" in the language of heterogeneous catalysis are now understood as "coordination" to the metal in homogeneous catalysis. The theory of heterogeneous catalysis is beginning to be described by many of the chemical terms[1] that originated in inorganic chemistry.

The task of distinguishing between these two major fields of catalysis has been made more ambiguous by the recent appearance of "heterogenized homogeneous catalysts." These are also called "homogeneous-heterogeneous catalysts," "polymer-supported homogeneous catalysts," or "polymer-anchored catalysts." These new catalysts are solid and insoluble, but the catalytically active centers are the same as those of homogeneous catalysts bound to the surface of polymers.

Immobilized enzymes also belong to this category. These new hybrid catalysts are treated as homogeneous catalysts because the chemical principles involved are the same as those of homogeneous catalysts.

Usually homogeneity in catalytically active sites is expected for a homogeneous catalyst. Homogeneity that is apparent at a glance does not ensure the presence of homogeneous active sites, however. Not all the ions in homogeneous catalysts are catalytically active. Sometimes only a few 'molar' percent of the catalyst is actively functioning as a catalyst under actual conditions.

Since the inhomogeneity of catalytic sites is frequently observed in homogeneous catalysis, these observations seem to reduce further the barrier between the two major fields of catalysis.

Homogeneous catalysts are often more selective, more active, and more reproducible, but are generally more difficult to remove after the reaction; they also are more vulnerable to extraneous materials, have shorter catalyst life, and are thermally more unstable when compared to usual heterogeneous catalysts.

A homogeneously catalyzed reaction can be analyzed by utilizing the usual chemical techniques. Kinetics and reaction energetics give information about the rate-determining steps just as in the case of stoichiometric reactions. Special techniques and theories as are used in analyzing heterogeneous reactions (particularly reactions catalyzed by solid catalyst surfaces) are *not* necessary in the analysis of homogeneous reactions. Some homogeneous catalysts are so robust that quenching the catalysts can be difficult because the catalysis sometimes continues during the work-up stage of the reaction. The possibility of artificial poisoning must be given all due consideration in these cases.

Mild homogeneous catalyst systems are easily examined along the reaction coordinate by nondestructive chemical analysis such as nmr, electron spin resonance (esr), and infrared (ir) spectroscopy using variable temperature measurements. Laser-excited Raman spectroscopy has recently been used in some cases, but its inherent low sensitivity could be greatly enhanced by utilizing resonance Raman techniques. Valuable and accurate insights into the structure of active metal complex catalysts can be provided by x-ray examination. Suitable combinations of these modern chemical techniques should aid in revealing more detailed mechanisms. These results can be further utilized to improve selectivity and activity of homogeneous catalysts.

The ease of investigation into the mechanism of homogeneous catalysis and the structure of homogeneous catalysts is a distinct advantage, opening rational ways to develop novel and intriguing catalysts.

2. SPECIFICITY AND SELECTIVITY*

Specificity and selectivity are well-known characteristics of homogeneous catalysis. The specificity of enzyme action is exemplified by the fact that only one enantiomer of lactic acid is dehydrogenated by L-lactic dehydrogenase. The enzyme is *specific* in reacting with L-lactic acid even in the presence of a large excess of D-lactic acid.

$$L(+)\text{-}CH_3\text{-}\underset{H}{\overset{OH}{C}}\text{-}CO_2H \xrightarrow{\text{L-lactic dehydrogenase}} CH_3\text{-}\underset{O}{\overset{\|}{C}}\text{-}CO_2H$$

$$D(-)\text{-}CH_3\text{-}\underset{OH}{\overset{H}{C}}\text{-}CO_2H \xrightarrow{\quad\text{//}\quad}$$

The probable structure of the dehydrogenation active center is as follows:

[Structure showing Prosthetic nicotinamide, His residue, and Arg residue interacting with the substrate]

---

* There has been some confusion concerning the definition and use of these terms. Textbooks of organic chemistry define the use of these terms in different ways. See E. L. Eliel, *Stereochemistry of Carbon Compounds,* McGraw-Hill, New York, 1962, pp. 434–446; J. B. Hendrickson, D. J. Cram, and G. S. Hammond, *Organic Chemistry,* 3rd ed., McGraw-Hill, New York, 1970. We define "specificity" as a very high degree of selectivity, for example, more than 95% selective.

Experiments using isotopic labels have shown that most enzymatic reactions proceed with high stereospecificity even when achiral substrates and achiral products are involved. For example, the enzyme-catalyzed redox reaction involving pyridinium-dihydropyridine systems has been found to proceed by addition-elimination of hydride from only one side of the pyridine nucleus.[2]

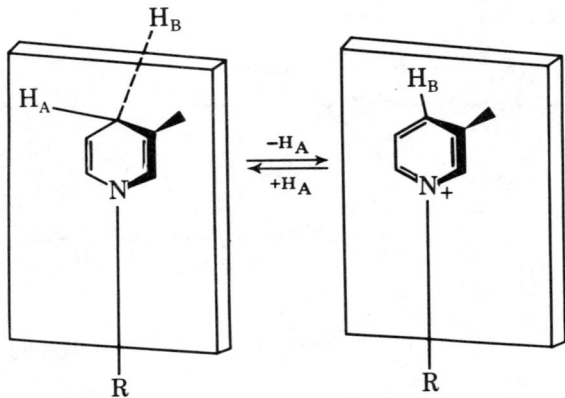

High specificity is not confined to enzymatic catalysis. The recent development of highly enantioselective* catalysts having chiral metal complexes has achieved enantioselective catalytic hydrogenation or hydrosilation with better than 90% selectivity.

* "Enantioselective means selective formation of one enantiomer in excess."

Selected Readings 13

Since enantioselectivity is considered to be the most difficult kind of selectivity to achieve in chemical reactions, this example suffices to demonstrate the high selectivity of artificial homogeneous catalysts in the present state of the science.

SELECTED READINGS

R. Ugo, *Catalysis Rev.*, *11*, 225 (1975).

B. Vennesland, "Stereoselectivity in Biology," in *Topics in Current Chemistry*, Vol. 48, Springer, Berlin, 1974.

G. C. Bond, *Heterogeneous Catalysis*, Clarendon Press, Oxford, 1974.

References for Chapter 2

1. See, M. Boudart, *Chem. Tech.*, *4*, 748 (1974); H. H. King, R. J. Pellet and R. J. Burwell, Jr., *J. Am. Chem. Soc.*, *98*, 5603 (1976).
2. K. E. Taylor and J. B. Jones, *J. Am. Chem. Soc.*, *98*, 5689 (1976).

3

BASIC PRINCIPLES

1. CHEMICAL KINETICS AND ENERGETIC ASPECTS

Homogeneous catalysis proceeds in solution just like an ordinary stoichiometric solution reaction, and chemical kinetics is one of the most useful and important experimental methods for studying both processes. Kinetics involves measuring the rates of a reaction under a variety of conditions, such as variation of temperature and of concentrations of reactants and catalysts. The rate of a reaction can be measured by monitoring the formation of a product or products or the disappearance of one or more starting materials. For meaningful kinetic results, the reaction in question must be reproducible and must yield definite products. A differential method (change of concentration per unit time) or an integral method (change of conversion or consumption after the reaction commences) is generally used to measure the rate. The measured rates are mathematically treated by rate equations (or rate laws). A variety of rate equations have been developed, but only a few simple ones are presented here.

1.1. Zeroth-Order Reactions

In a zeroth-order reaction the amount of product formed is directly proportional to reaction time and does not depend on the concentration of any of the reactants.

$$\frac{d[\text{prod}]}{dt} = k$$

[prod] = concentration of the product(s)

This kinetic behavior has been found in some catalytic and photochemical reactions, especially in catalytic reactions having a stable substrate-catalyst complex. If the product is formed from the rate-determining unimolecular decomposition of this complex, the rate depends only on the concentration, which is constant, of the complex.

1.2. First-Order Reactions

In a first-order reaction the rate depends on the concentration of one of the reactants. A linear relationship holds between the logarithm of reactant concentration at any time (t)/initial reactant concentration and time t. Letting $[\text{react}]_0$ be the initial concentration of the reactant and $[\text{react}]$ the concentration of the reactant at time t, the following equation can be written:

$$\text{rate} = -\frac{d[\text{react}]}{dt} = k_1[\text{react}]$$

$$-2.303 \log\left(\frac{[\text{react}]}{[\text{react}]_0}\right) = k_1 t$$

Usually $[\text{react}]_0$ is symbolized as a and $[\text{react}]$ as $a - x$, where x represents amount reacted at any particular time.

The first-order rate constant k_1 can be obtained from the slope of the straight line obtained when t is plotted versus $2.303 \log([\text{react}]/[\text{react}]_0)$.

In a first-order reaction the half-life $t_{1/2}$ of a reactant is related to the first-order rate constant k_1 in the following manner:

$$t_{1/2} = \frac{2.303}{k_1} \log 2 = \frac{0.693}{k_1}$$

Therefore measurement of the half-life also leads to k_1.

The homogeneously catalyzed cis-trans isomerization of azobenzene serves as an example of a reaction with first-order kinetics.[1]

The rate can be measured continuously by monitoring the visible absorption at 510 nm, and it is exactly proportional to concentration of remaining *cis*-azobenzene. The apparent activation energy (14.7 kcal/mol) of the catalyzed reaction is considerably lower than that (~22 kcal/mol) of the uncatalyzed thermal isomerization.

1.3. Pseudo-First-Order Reactions

Pseudo-first-order reactions have one component in great excess. In such cases rates do not depend on the concentration of the excess component (because it changes by a relatively small amount), and one can measure the dependence of the rate on the concentration of the other component only. Therefore for a fixed catalyst concentration, a second-order reaction becomes practically first-order. This experimental technique gives good results when the active catalyst is formed (or is protected from deterioration) by weak interaction with the reactants(s) used in excess.

It is important to follow the reaction as long as possible to distinguish the order of the reaction. In catalytic reactions, however, this is not always easy, because some catalysts exhibit induction periods, and others tend to decompose during the course of the reaction. One must investigate catalysis rigorously to obtain reliable kinetic data, otherwise the result will not be of scientific value.

1.4. Complex Reactions

Homogeneous catalysis in many cases involves a complex series of elementary reactions, and the order of such catalytic reactions generally is not simple. Fractional orders (e.g., 1.2) have frequently been found. In addition, the presence of an induction period and the deterioration of an active catalyst have major affects on the rate profile. Figure 5 gives time versus conversion curves for typical complex reactions.

As an example of a complex reaction, let us consider a reaction between A and B catalyzed by K. We assume that at first A and K interact to give an intermediate product AK, which reacts with B to give the product C via another intermediate ABK. This is a consecutive reaction.

The time-conversion curve in the initial stage of the reaction cannot be analyzed simply. When the reaction is proceeding steadily, the concentration of the intermediate becomes constant. In this "stationary state" (steady state), a rate equation can be set up as follows, utilizing the simplifying assumption that the

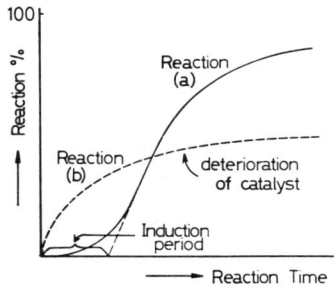

Fig. 5. Reactions involving an induction period and catalyst deterioration.

concentration of intermediate does not change with time. That is,

$$\frac{d[AK]}{dt} = \frac{d[ABK]}{dt} = 0$$

The reaction scheme is as follows:

$$A + K \underset{k_2}{\overset{k_1}{\rightleftarrows}} AK$$

$$AK + B \underset{k_4}{\overset{k_3}{\rightleftarrows}} ABK \overset{k_5}{\rightarrow} K + C$$

The rate equations based on the assumption of stationary states are:

$$\frac{d[ABK]}{dt} = k_3[AK][B] - (k_4 + k_5)[ABK] = 0$$

$$\frac{d[AK]}{dt} = k_1[A][K] - k_2[AK] - k_3[AK][B] + k_4[ABK] = 0$$

These are combined to give:

$$[AK] = \frac{k_1[A][K]}{k_2 + \left(k_3 - \dfrac{k_3 k_4}{k_4 + k_5}\right)[B]}$$

Now the rate equation is:

$$\frac{d[C]}{dt} = k_5[ABK] = \frac{k'[A][B][K]}{k'' + (1-k)[B]}$$

where

$$k = \frac{k_4}{k_4 + k_5}$$

$$k' = \frac{k_1 k_5}{k_4 + k_5}$$

$$k'' = \frac{k_2}{k_3}$$

As expected, the equation contains too many rate constants to be easily verified experimentally. When $k'' \gg 1$, that is, $k_2 \gg k_3$ (concentration of [AK] is very small), the equation simplifies to

$$\frac{d[C]}{dt} = \frac{k'}{k''} [A][B][K]$$

This equation shows that the rate dependence is first order for each reactant and for the catalyst. The rate-determining step in this case is the reaction to form the activated thermolecular intermediate [ABK]. In contrast, when $0 < k'' \ll 1$ ($k_2 \ll k_3$), k'' is negligible compared to $(1 - k)[B]$. The rate equation then becomes

$$\frac{d[C]}{dt} = \frac{k'}{(1 - k)} [A][K]$$

This indicates that the reaction of AK with B is fast and decomposition of ABK to C and K is also fast. Now the rate-determining step is the reaction of A with K. For example, hydrogenation with $RhH(CO)(PPh_3)_3$ has been reported to have a similar kinetic equation.[2]

$$\text{rate} = \frac{k K[H_2][\text{olefin}][RhH(CO)(PPh_3)_2]}{1 + K[\text{olefin}]}$$

The rate constant k and the equilibrium constant K are derived from the following equation.

$$RhH(CO)(PPh_3)_2 + \text{olefin} \underset{}{\overset{K}{\rightleftarrows}} Rh(\text{alkyl})(CO)(PPh_3)_2$$

$$Rh(\text{alkyl})(CO)(PPh_3)_2 + H_2 \overset{k}{\rightarrow} RhH(CO)(PPh_3)_2$$

1.5. The Michaelis-Menten Rate Equation

The Michaelis-Menten rate equation was developed to characterize the variation of rates observed in enzymic reactions and has been utilized frequently in describing homogeneously catalyzed reactions. The Michaelis-Menten rate equation is suitable when the reaction intermediate is reversibly formed from the reactants and the catalyst (or enzyme) and is relatively kinetically stable with respect to irreversible decomposition to products. Some part of the catalyst (or enzyme [E]) therefore exists as the intermediate (or enzyme-substrate (ES) complex) and accumulates in some reactions (i.e., $[E_0] = [E] + [ES]$). The situation is represented by the following simple scheme:

$$E + S \underset{k_{-1}}{\overset{k_1}{\rightleftharpoons}} ES \overset{k_2}{\rightarrow} E + P$$

where E = enzyme (catalyst)
S = substrate (reactant)
P = product

The rate after the steady state of the reaction is established can be expressed as described previously and yields the following rate equation:

$$\text{rate} = \frac{k_2[E][S]}{\left(\dfrac{k_2 + k_{-1}}{k_1}\right) + [S]} = \frac{V_{max}[S]}{K_m + [S]}$$

In the presence of a large amount of substrate the rate approaches V_{max} (maximum velocity), which is determined only by the amount of enzyme present and the reaction conditions. The K_m value (Michaelis constant) gives the substrate concentration at which the rate is one-half the V_{max} value.

The k_2 value is related to the turnover number and the value indicating the activity of the enzyme concerned. To obtain V_{max} and K_m, reciprocal rate is plotted against reciprocal substrate concentration (a Lineweaver-Burk plot), and these values can be obtained as shown in Fig. 6b.

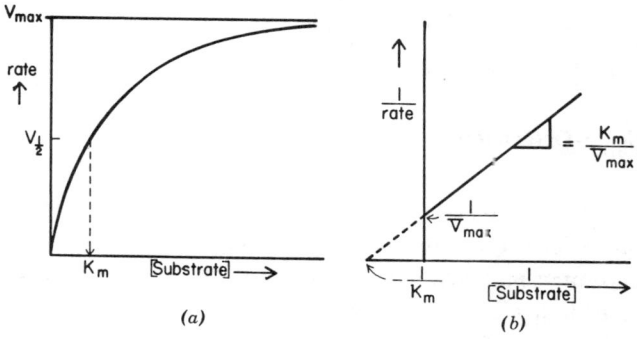

Fig. 6. Illustration of the relationship between rates and substrates. (a) Rate versus [substrate] and (b) reciprocal rate versus reciprocal [substrate].

The following complex rate equation has been found for olefin hydrogenation with RhCl(Ph$_3$P)$_3$ ("Wilkinson's catalyst"). A reaction scheme has been developed to fit the rate equation[3]:

$$\text{rate} = \frac{k_1 k_1 [\text{H}_2][\text{active cat.}][\text{olefin}]}{1 + K_1[\text{H}_2] + K_2[\text{olefin}]}$$

$$(\text{active cat.}) + \text{H}_2 \xrightleftharpoons{K_1} (\text{active cat.}\cdot\text{H}_2)$$

olefin $\updownarrow K_2$ olefin $\downarrow k_1$

$$(\text{active cat. olefin}) \xrightarrow{\text{H}_2} (\text{active cat.}) + \text{alkane (product)}$$

1.6. Rate Dependence on Catalyst Concentration

We have assumed that the catalyst concentration is constant throughout the reaction, therefore does not appear as a variable in the rate equation. Experiments with several different catalyst concentrations are necessary to understand the catalysis process. The dependence of rate on catalyst concentration may be complex because some fraction of the catalyst may be poisoned by impurities or because the catalyst may decompose after catalyzing the reaction of several molecules. A recent actual example of the rate dependence of the reaction on catalyst concentration for the following reaction is shown in Fig. 7.[4]

$$\text{1-hexene} \xrightarrow[\text{RhH(PR}_3)_3]{\text{H}_2} \text{hexane}$$

Also, the dependence of rate on catalyst concentration can result from phosphine ligand dissociation.[4]

$$\text{HRhP}_3 \rightleftharpoons \text{HRhP}_2 + \text{P}$$

where P = phosphine ligand

1.7. The Rate-Determining Step

In general the kinetic rate equation tells us which step is the slowest (rate-determining step). The steps preceding the slowest step are sometimes denoted as preequilibrium or prereaction steps. Nothing however, can be learned about the faster steps following the rate-determining step. The following examples

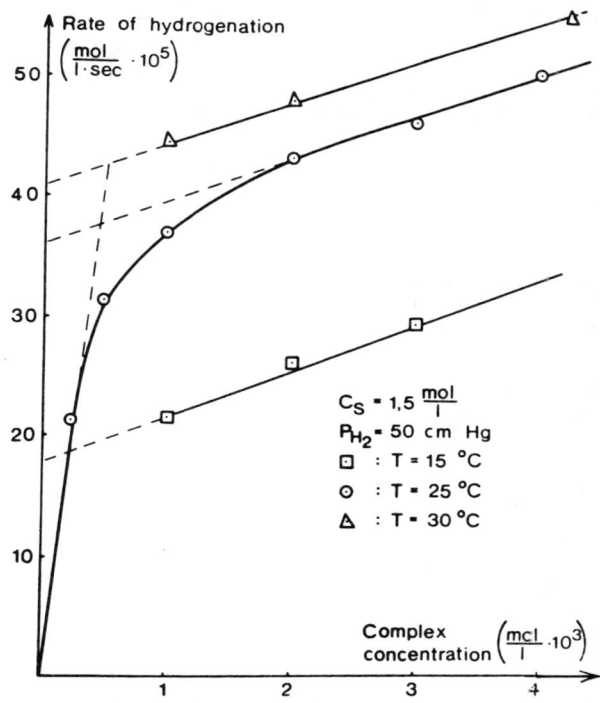

Fig. 7. Dependence of the rate of the catalyzed reaction on the catalyst concentration: a practical case (olefin hydrogenation).

of preequilibrium steps are for (1) olefin-induced reductive elimination of

(a) $Ni(Et)_2L_2 + CH_2{=}CHR \rightleftharpoons Ni(Et)_2L_2(CH_2{=}CHR)$

$$Ni(Et)_2L_2(CH_2{=}CHR) \xrightarrow[\text{the rate-determining step}]{} Ni(CH_2{=}CHR)L_2 + Et{-}Et$$

(b) $Ni(CN\text{-}t\text{-}Bu)_4 \rightleftharpoons Ni(CN\text{-}t\text{-}Bu)_3 + t\text{-}BuNC$

$$2Ni(CN\text{-}t\text{-}Bu)_3 + 2O_2 \xrightarrow[\text{the rate-determining step}]{} Ni(CN\text{-}t\text{-}Bu)_3O_2 + Ni(CN\text{-}t\text{-}Bu)_2O_2 + t\text{-}BuNC$$

$NiEt_2L_2$ producing butane and (2) oxidative addition of O_2 to $Ni(CN\text{-}t\text{-}Bu)_4$.

1.8. Activation Parameters

When the rate constant is measured at several different temperatures, one obtains additional information about the reaction. First, from the Arrhenius equation one can obtain the activation energy for the reaction. The logarithms

$$k_1 = A \exp\left(\frac{-E_a}{RT}\right)$$

of the first-order rate constant k_1 are plotted against the reciprocal of the absolute temperature T to give a straight line in a short temperature range (usually $\pm 20°$). The slope of the line yields the activation energy E_a and the intercept yields the "frequency factor" A. The value of A can yield the degree of randomness in the transition state.

$$\ln A - \ln k_1 = \frac{E_a}{RT}$$

From the equation above and by utilizing the following equations from transition state theory where k is Boltzmann's constant and h is Planck's constant, one can obtain values for $\Delta H^{\ddagger}$ and $\Delta S^{\ddagger}$, the enthalpy and entropy of activation, respectively.

$$k_1 = \frac{kT}{h} \exp\left(\frac{-\Delta H^{\ddagger}}{RT}\right) \exp\left(\frac{\Delta S^{\ddagger}}{R}\right)$$

$$E_a = \Delta H^{\ddagger} + RT$$

$$\Delta S^{\ddagger} = R \ln \frac{Ah}{kT}$$

If the measured k refers to a single step, not to a combination of rate constants and (or equilibrium constants), these values, sometimes called kinetic parameters, gives an important insight into the mechanism of a reaction. Uncatalyzed reactions usually have high activation enthalpies. When the same reaction is catalyzed, the energetic barrier (the activation enthalpy) becomes generally lower. Kinetic parameters for some typical examples are shown in Table 1.

As a further example, the following kinetic parameters were obtained for the catalytic hydrogenation of cyclohexene in a benzene-hexane solution at $25°$ by Wilkinson's catalyst (see also the scheme on p. 20).[5]

$$k_1 = 0.15 M^{-1} \text{sec}^{-1} \qquad K_2/K_1 \sim 1.6 \times 10^{-3}$$

$$\Delta H^{\ddagger} = 22.3 \text{ kcal/mol}; \qquad \Delta S^{\ddagger} = +12.9 \text{ eu}$$

Table 1. Examples of Activation Energy: A Comparison Between Catalyzed and Uncatalyzed Reactions[6]

Reactions	Catalyst	Activation Energy E_a (kcal)
Decomposition of H_2O_2	None	18.0
	Colloidal platinum	11.7
	Liver catalase	5.5
Hydrolysis of $C_3H_7CO_2C_2H_5$	H^+	13.2
	Pancreatic lipase	4.5
Hydrolysis of $CO(NH_2)_2$	H^+	24.6
	Urease	12.5

Reactions with low activation enthalpies can proceed at much lower temperatures than the corresponding uncatalyzed reactions. Since the temperature dependence of rate in a catalyzed reaction is small, the catalyzed reaction can be performed over a wider temperature range than that of an uncatalyzed reaction (Fig. 8). This characteristic is a distinct advantage in controlling catalytic reactions.

Activation entropies are also important because large negative values of $\Delta S^\ddagger$ (-10 to -30 eu) are generally associated with a bimolecular rate-determining step between nonionic reactants, in which two molecules come together to form the transition state. For example, in reactions where the rate-determining step is coordination of a molecule to a metal, a $\Delta S^\ddagger$ value of -20 eu can be expected. In contrast, reactions with dissociative transition states usually give positive $\Delta S^\ddagger$ values ($\Delta S^\ddagger \sim +5$ eu).[7]

When $\Delta S^\ddagger$ values are large, the height of the free energy barrier at the rate-determining step ($\Delta G^\ddagger$) changes with temperature. Thus reactions proceeding

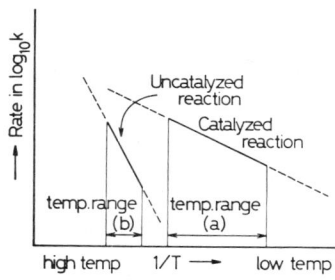

Fig. 8. Arrhenius diagrams for catalyzed and uncatalyzed reactions.

by coordination and/or dissociation sometimes occur through different pathways depending on the temperature. For example, the ligand replacement reaction of $TaBr_5 \cdot Et_2O$ with Me_2S proceeds by an associative mechanism at $-8°$ but occurs through a dissociative path above $52°$.[8]

$$TaBr_5 \cdot Et_2O \underset{}{\overset{>50°}{\rightleftharpoons}} \{TaBr_5\} \underset{}{\overset{Me_2S}{\rightleftharpoons}} TaBr_5Me_2S + Et_2O$$
$$\text{Dissociative path}$$

$$\overset{-10°}{\longrightarrow} \{TaBr_5 \cdot Et_2O \cdot Me_2S\}$$
$$\text{Associative path}$$

Generally dissociative paths with positive $\Delta S^\ddagger$ are favored at higher temperatures because the term $-T\Delta S^\ddagger$ becomes larger, resulting in the lowering of the energy barrier.

2. HOMOGENEOUS ACTIVE SITES: ACTIVATION AND DEACTIVATION

Heterogeneity at the active sites is one of the most important properties of heterogeneous catalysis. Similarly, in homogeneous catalysis every molecule of catalyst has nominally been considered to be an active site. However closer examination of actual homogeneous catalysts has revealed that only a part of the catalyst species is active in many cases. For example, the catalytically active species of Wilkinson's hydrogenation catalyst $RhCl(PPh_3)_3$ has been identified as a labile $RhCl(PPh_3)_2(S)$ species, where S is a solvent (e.g., C_6H_6) or a weakly coordinating olefin. $RhCl(PPh_3)_3$ itself is not active if no dissociation of PPh_3 is taking place. The catalyst is also activated by a small amount of oxygen that oxidizes dissociated free PPh_3 to Ph_3PO to enhance the formation of the active species $RhCl(PPh_3)_2(S)$.

Also the addition of excess tertiary phosphines, especially less bulky trialkylphosphine, deactivates the hydrogenation catalyst. Addition of such chelating diphosphines as $R_2PCH_2CH_2PR_2$ (R = Ph, Et), also deactivates the catalysts.

$$RhCl(PPh_3)_3 \underset{+PPh_3}{\overset{-PPh_3}{\rightleftharpoons}} RhCl(PPh_3)_2 \overset{S}{\rightleftharpoons} RhCl(PPh_3)_2(S)$$
$$\text{Very unstable} \qquad \text{Active species}$$

$$PR_3 \updownarrow \qquad\qquad H_2 \updownarrow$$

$$RhCl(PPh_3)_2(PR_3) \qquad RhH_2Cl(PPh_3)_2(S)$$
$$R = Et \text{ or } CH_2Ph$$

Most low-valent metal complex catalysts are generally deactivated by air and in some cases by water. Nickel(0)-ligand catalysts, Ziegler-type catalysts, and metal hydride catalysts (e.g., MoH_2Cp_2) are readily deactivated (or poisoned) by air. Careful purification of the reactants and the solvents is usually required for catalysis to occur. Carbon monoxide, hydrogen cyanide, and PH_3, frequently act as poisons for these kinds of catalyst, and it is well known that these molecules also strongly deactivate many heterogeneous catalysts.

The phenomenon of poisoning by strongly coordinating molecules is due to formation of catalytically inert complexes. For example, the following reactions explain the poisoning of Wilkinson's catalysts for olefin hydrogenation.

$$RhCl(PPh_3)_3 \xrightarrow{CO} RhCl(CO)(PPh_3)_2$$
$$\downarrow PH_3$$
$$RhCl(PH_3)_n(PPh_3)_{3-n}$$

Addition of an appropriate donor molecule to catalyst systems sometimes activates the catalysis. Thus an equimolar amount of pyridine or similar σ bases to cobaloxime [$Co(dmg)_2$, dmg = dimethylglyoximato] catalyst has been known to enhance hydrogenation activity.[9]

$$PhCOCOPh \xrightarrow[Co(dmg)_2/py]{H_2} Ph\text{-}\underset{\underset{OH}{|}}{C}H\text{-}COPh$$

In principle, the effect of these coordinating additives is explained by coordination to the active metal center. However it is not always possible to predict the effect because of lack of knowledge of the catalytic species.

3. FUNDAMENTAL ASPECTS OF SELECTIVITY

Selectivity is an important characteristic in any chemical reaction, especially in a catalytic reaction. There are several different types of selectivity. Substrate selectivity—the ability to react with one compound in preference to another—is common to virtually all catalysts, but is most prevalent for enzymes. Substrate selectivity is a difficult subject to cover, however, because it can be either relatively simple as with molecular homogeneous catalysts, or extremely complex, as in the case of enzymes. The distinction between specificity and selectivity has in many cases remained ambiguous. For example, the following catalytic and stoichiometric reactions have been called "stereospecific" or "stereoselective":

$$CH_3-C\equiv C-CH_3 \xrightarrow{H_2/Pd-C\ catalyst} \begin{array}{c} CH_3 \\ \diagdown \\ H \end{array} C=C \begin{array}{c} CH_3 \\ \diagup \\ H \end{array}$$

$$\xrightarrow{Na/liq\ NH_3} \begin{array}{c} CH_3 \\ \diagdown \\ H \end{array} C=C \begin{array}{c} H \\ \diagup \\ CH_3 \end{array}$$

$$\begin{array}{c} H \\ \diagdown \\ HO_2C \end{array} C=C \begin{array}{c} CO_2H \\ \diagup \\ H \end{array} + \begin{array}{c} H \\ \diagdown \\ H \end{array} C=C \begin{array}{c} CO_2H \\ \diagup \\ CO_2H \end{array} \xrightarrow[aspartase]{NH_3} \begin{array}{c} NH_2 \\ H \diagdown | \\ C \\ | \\ CH_2-CO_2H \end{array} CO_2H$$

Recently several new chemical terms have been proposed to describe rationally the selectivity in organic reactions. "Regioselectivity," "enantioselectivity," and "diastereoselectivity" are among these new terms. Here some typical examples are explained. The first example consists of a regioselective catalytic reaction:

$$RCH=CH_2 + H_2 + CO \xrightarrow{[Co(CO)_3(PR_3)]_2} R-CH_2-CH_2-CHO + R-CH-CH_3$$
90% regioselectivity

$$\begin{array}{c} | \\ C=O \\ | \\ H \end{array} \quad 10\%$$

$Co_2(CO)_8$ also catalyzes the same reaction, but the amount of branched aldehyde formed increases to 20 to 30%. Replacing one or more (CO) ligands with a suitable phosphine ligand raises the regioselectivity of the catalyzed addition of the CHO group to the terminal olefinic carbon. Regioselectivity toward the branched aldehyde can be promoted by using $Rh_2(CO)_8$ instead of $Co_2(CO)_8$.[10]

$$RCH=CH_2 + H_2 \xrightarrow{Rh_2(CO)_8} \begin{array}{c} RCH-CH_3 \\ | \\ CHO \end{array} \quad \sim 50\%$$

$$+$$

$$R-CH_2-CH_2-CHO \quad \sim 50\%$$

Fundamental Aspects of Selectivity

When an asymmetric carbon (chiral carbon) is generated by the addition of a molecule X—Y across the double bond as shown below, the reaction can be *enantioselective* toward the *prochiral* unsaturation. These prochiral olefins or ketones have a pair of *enantiofaces*. If there is any selection between the enantiofaces, the addition is called an *enantioface differentiation reaction* or simply an "*enantioselective reaction.*"

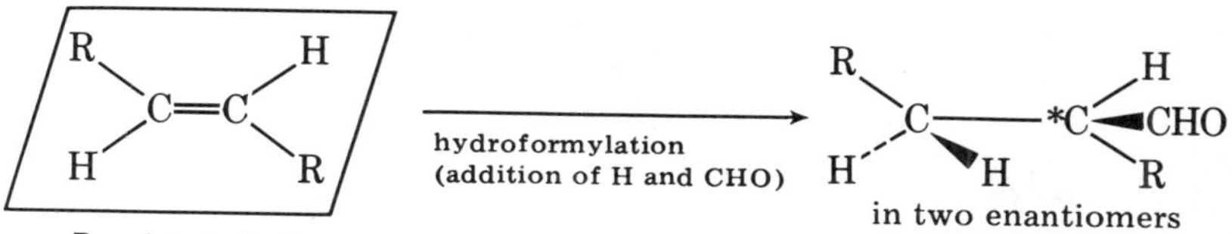

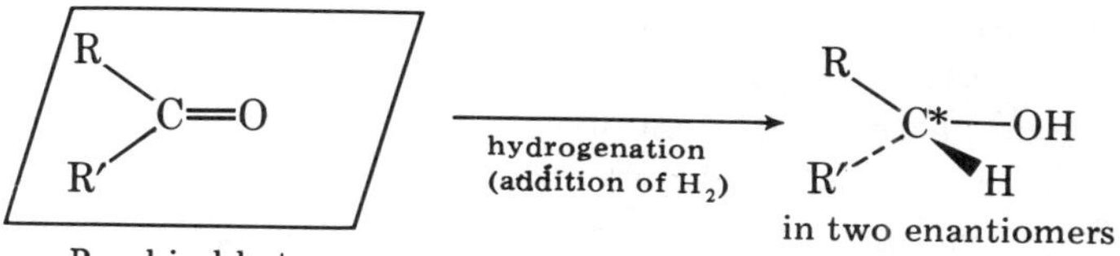

Cyclopropanation of prochiral olefins can also be made enantioselective. It is of interest to point out that alkyl diazoacetates (N_2CHCO_2R) are also prochiral molecules and yield an asymmetric carbon on cyclopropanation even with nonprochiral olefins. Therefore cyclopropanation of prochiral olefins with prochiral diazo compounds gives two pairs of enantiomeric products.

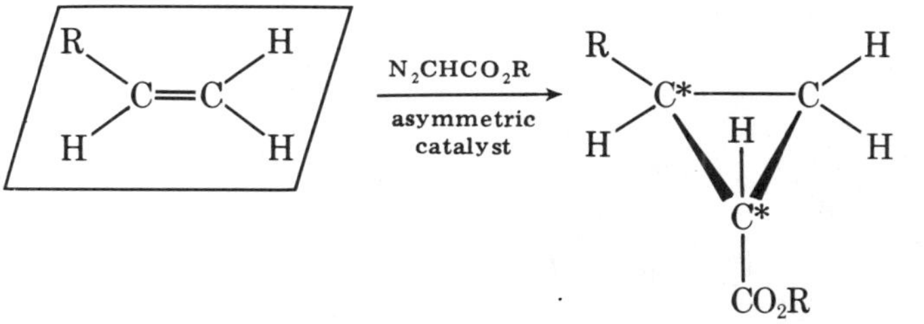

Chiral catalysts, in principle, can select or differentiate between the enantiofaces, and suitable chiral catalysts can perform enantioselective addition to prochiral unsaturation to form one enantiomer in excess. Thus a small amount of an efficient enantioselective catalyst can produce a large amount of optically active organic compounds from achiral reactants. Some chiral catalysts interact more strongly with one enantiomer out of a racemic enantiomer mixture. When one enantiomer (e.g., or propylene oxide) is selectively decomposed to achiral products (e.g., acetone), the other enantiomer is left unchanged. Such selection is called "kinetic control." [11]

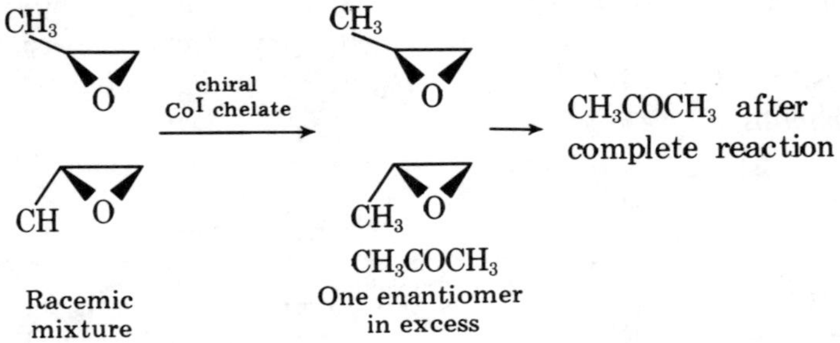

Racemic mixture → chiral Co^I chelate → One enantiomer in excess CH$_3$COCH$_3$ → CH$_3$COCH$_3$ after complete reaction

When an achiral carbon atom becomes asymmetric through a substitution reaction, the reaction can be made enantioselective by utilizing chiral catalysts. For example, the hydroxylation of palmitic acid occurs under biochemical conditions to give enantiomeric 2-hydroxypalmitic acid in large excess.[12] The two α-hydrogen atoms in palmitic acid are *enantiotopic* and each is called pro-*R* or pro-*S* by application of the sequence rule* as shown below for CH$_2$ hydrogens of ethanol.

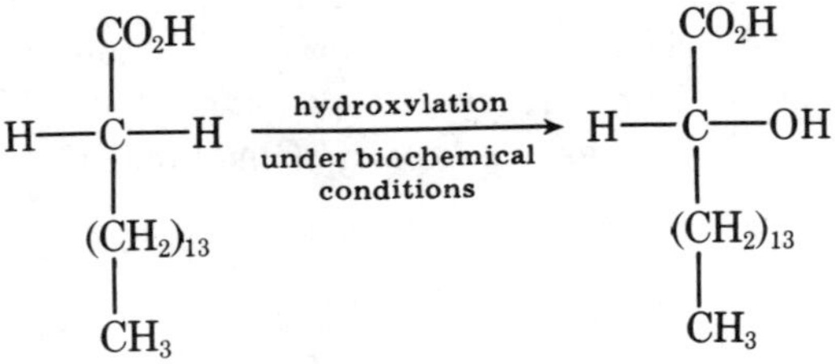

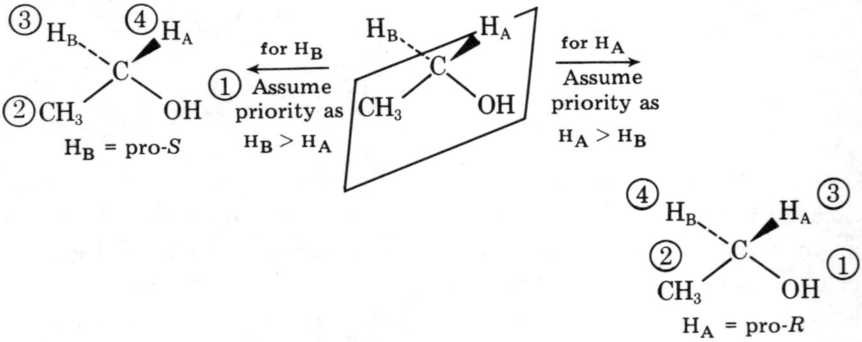

H—C(CO$_2$H)(CH$_2$)$_{13}$(CH$_3$)—H $\xrightarrow{\text{hydroxylation under biochemical conditions}}$ H—C(CO$_2$H)(CH$_2$)$_{13}$(CH$_3$)—OH

③ H$_B$, ④ H$_A$, ② CH$_3$, OH
H$_B$ = pro-*S*

$\xleftarrow{\text{for H}_B}$ ① Assume priority as H$_B$ > H$_A$

H$_B$, H$_A$, CH$_3$, OH

$\xrightarrow{\text{for H}_A}$ Assume priority as H$_A$ > H$_B$

④ H$_B$, ③ H$_A$, ② CH$_3$, ① OH
H$_A$ = pro-*R*

* Priority in the sequence rule is H$_A$ > H$_B$ for H$_A$ in the determination of their priority and H$_B$ > H$_A$ for H$_B$.

Fundamental Aspects of Selectivity

Alcohol dehydrogenase differentiates these prochiral hydrogens, and ethanol is dehydrogenated to acetaldehyde by an enantioselective H-abstraction reaction. The same enzyme also catalyzes the hydrogenation of acetaldehyde in an enantioselective way. In this case, of course, the selectivity is not observable without isotopic substitutions, since both molecules are not chiral.

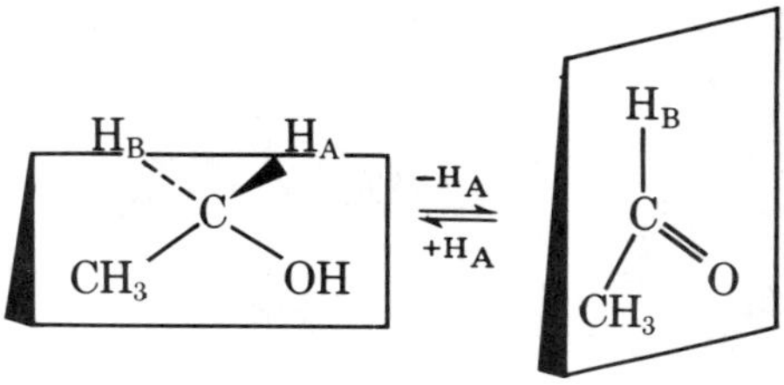

Diastereoselective reactions deal with the stereoselectivity of reactions on diastereofaces or at diastereotopic sites. Typical examples of diastereofaces are shown below:

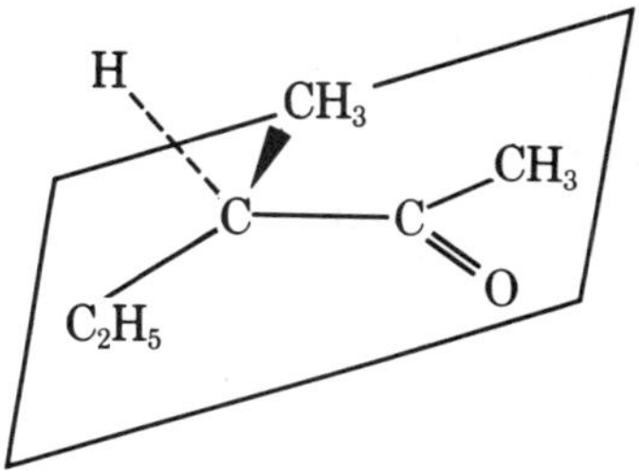

Diastereofaces of
C_2H_5CH—$COCH_3$
|
CH_3

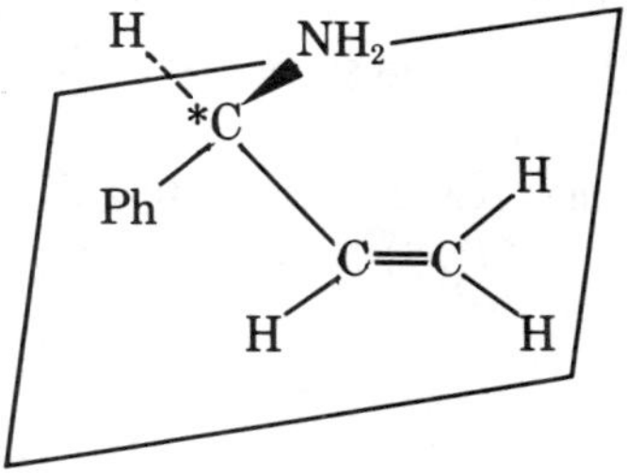

Diastereofaces of
$PhCH$—CH=CH_2
|
NH_2

Addition to an olefin or carbonyl (C=C or C=O) on a molecule containing diastereofaces can create one or more new asymmetric centers (asymmetric carbons), resulting in the formation of a diastereomers.

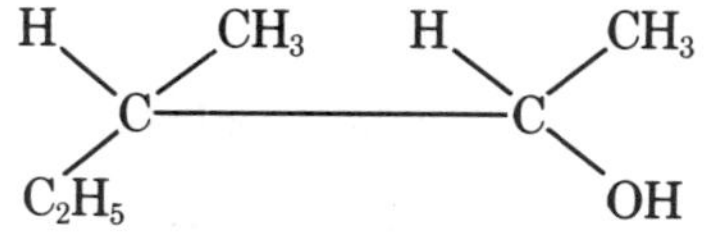

Selective formation of one particular diastereomer is called diastereoselection. For example, an optically active ferrocenyl aldehyde (see below) yields the corresponding carbinol by methylation with MeMgI with 100% diastereoselectivity.[13]

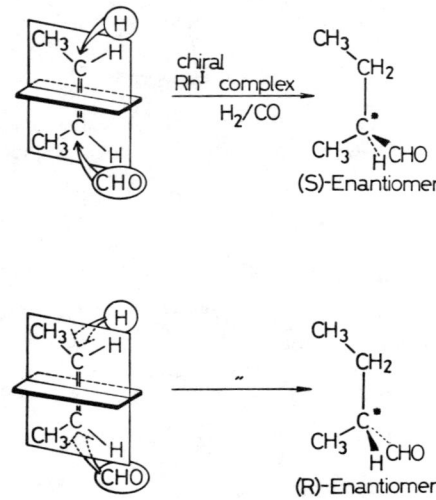

(R)-(−)-Isomer → (R, R)-(−)-Isomer, 100% optical purity

The two hydrogens of a methylene group linked to a chiral center are called diastereotopic hydrogens and are nonequivalent. This nonequivalence can be revealed by the ¹H nmr spectrum exhibiting an AB-type nmr pattern in most cases. Ordinary *achiral* reagents or catalysts can distinguish between these hydrogens, and in some cases selective replacement can occur. This process is called diastereoselection.

When a reagent can distinguish between the right- or left-hand side of a molecule (Fig. 9), an enantioselective reaction can occur even if there are no

Fig. 9. An example of selective formation of one enantiomer by a chiral catalyst.

Fundamental Aspects of Selectivity

enantiofaces. If the rate of addition of the H and CHO groups differs between these two reactions as influenced by the chirality of the catalyst, one enantiomer will be formed in preference (see Chapter 5, Section 14).

In some cases regioselection and enantioselection work cooperatively to give one enantiomer in excess (see below).

$$CH_3-CH_2 \quad H$$
$$\diagdown\diagup$$
$$C$$
$$\|$$
$$C$$
$$\diagup\diagdown$$
$$CH_3 \quad H$$

$\xrightarrow[\text{chiral catalyst}]{H_2/CO}$ (with regioselection)

$$CH_3-CH_2$$
$$\diagdown$$
$$CH_2$$
$$|$$
$$*CH$$
$$\diagup\diagdown$$
$$CH_3 \quad CHO$$

Chiral product (usually optically active)

$\xrightarrow[\text{regioselective catalyst}]{H_2/CO}$ (selective to CHO addition to the inner olefinic carbon)

$$CH_3-CH_2$$
$$\diagdown$$
$$CH-CHO$$
$$\diagup$$
$$CH_3-CH_2$$

Achiral product (optically inactive)

The extent of enantioselection can be expressed by the *optical purity* of the product, which is based on the optical activity of the product and the known optical rotation of the pure material.

$$\text{optical purity} = \frac{\text{observed optical rotation } [\alpha]}{[\alpha] \text{ of optically pure product}} \times 100$$

It is also important to examine the optical purity of the chiral catalyst by measuring the optical activity of the chiral ligand or, in some cases, of the chiral complex itself. Some organic compounds (e.g., chiral alkanes) have very small values of optical rotation even in high optical purity, making it very difficult sometimes to measure optical purity with any accuracy.

The enantiomeric purity of an organic acid can also be measured by formation of the corresponding diastereomeric esters with optically pure chiral alcohol (e.g., *l*-menthol) followed by separation of the two esters by gas liquid chromatography (glc). Recently nmr techniques have been utilized to measure the ratio of enantiomers by "optishift reagents" or by formation of diastereomers with suitable optically pure reagents. The ratio (e.g., $[(R)]/[(S)]$) obtained this way can be usually expressed as the *percentage of enantiomeric excess* (% ee) as is described below:

$$\% \text{ ee} = \frac{[(R)] - [(S)]}{[(R)] + [(S)]} \times 100$$

The enantiomeric excess is the same as the optical purity in most cases. In special cases (e.g., $HO_2CCH_2-(CH_3)(Et)C^*-CO_2H$), these two values have been found to be quite different because of selective association between (R)- and (S)-isomers in solution (A. Horeau, *Tetrahedron Lett., 1969,* 3121). When enantioselection occurs in the rate-determining step, the free energy difference between the transition states leading to (R)- and (S)-isomers has been correlated to the % ee value by the following equation derived from the fact that this selection is kinetically controlled (Curtin-Hammett principle).[14]

$$\Delta\Delta G^{\ddagger}_{R-S} = -RT \ln \frac{(100 + ee)}{(100 - ee)}$$

The free energy difference $\Delta\Delta G^{\ddagger}$ is illustrated in Fig. 10a, and the $\Delta\Delta G^{\ddagger}$ required to produce a given enantiomeric excess is shown in Fig. 10b.

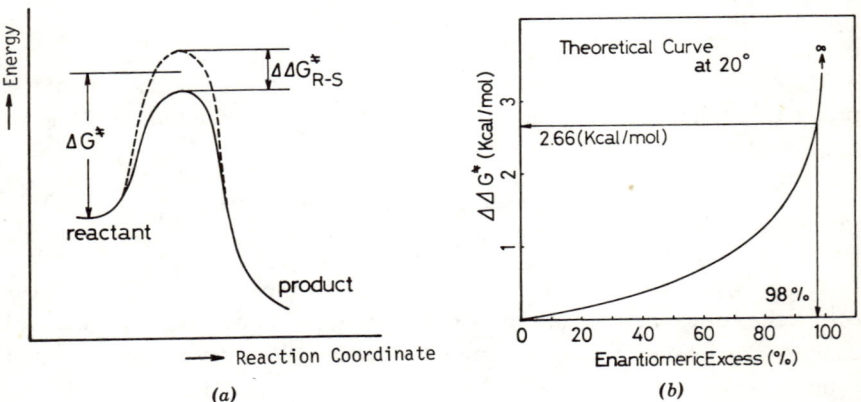

Fig. 10. (a) Energetic aspects of enanthioselective reactions. (b) The relation between $\Delta\Delta G^{\ddagger}$ and %ee values.

4. STEREOCHEMISTRY, ORBITAL SYMMETRY, AND REACTIVITY

The stereochemistry around a particular atom or ion is an important consideration in determining the reactivity of the chemical species in question. Symmetry properties of the relevant orbitals in such species are closely related to the stereochemistry, hence are related to its reactivity. Therefore the three basic factors of stereochemistry, orbital symmetry, and reactivity must all be considered in deriving the fundamental principles of homogeneous catalysis.

Let us begin with the simplest chemical species of all, H^+. The proton has a vacant $1s$ orbital that has a totally symmetric spherical spatial extension. Therefore no steric requirement for interaction with any σ-type donor exists. Boron trifluoride (BF_3) is a planar molecule of D_{3h} symmetry and has a vacant p_z orbital directed along the z (C_3) axis (Fig. 11a). Here the stereochemistry around the boron atom forces the reactive σ-accepting center to extend above and below the molecular plane. BF_3 is a strong σ acceptor, Lewis acid, and an electrophile because it attacks the electronic portions of other molecules in reactions. Ammonia (NH_3) has a trigonal pyramidal structure (C_{3v}), and a reactive, filled nonbonding nitrogen orbital (mainly a mixed orbital of s and p),

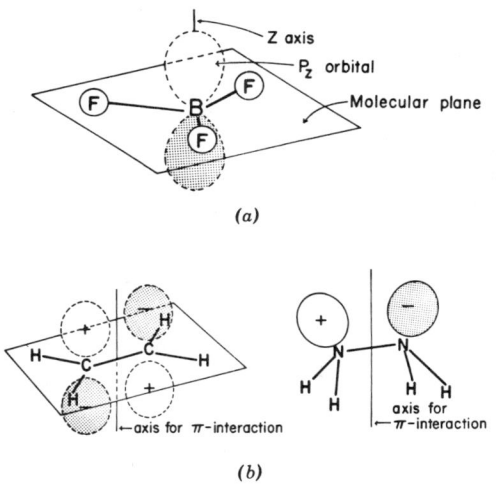

Fig. 11. (a) The electrophilic p_z orbital of BF_3. (b) The π-acidic orbital (π^*) of ethylene and the π-basic orbital of hydrazine.

which is directed along $+z$ axis. This axis is the most favorable direction for an entering electrophile. A typical σ acceptor such as BF_3 interacts strongly with ammonia to give a σ-donor-acceptor (σ-DA) complex. The stereochemistry around the nitrogen atom in NH_3 restricts electrophilic attack to the $+z$ axis.

As molecules become more complex in structure, the symmetry and shape of filled or vacant orbitals becomes accordingly more complex. The lowest vacant

π^* orbital of ethylene functions as a π acceptor (π acid) (Fig. 11b), whereas the two lone pairs on the nitrogen atoms in hydrazine can function as π donors (π bases) in the conformation shown in Fig. 11b.

It is now clear that suitable combinations of light elements in certain steric arrangements generate unique symmetries for the frontier orbitals* depending on electron occupancy and the nature of participating orbitals. For illustration, the tricoordinated carbon (sp^2) frontier orbitals assume unique symmetry on combination with other sp^2 carbon atoms in a suitable steric arrangement (Fig. 12).

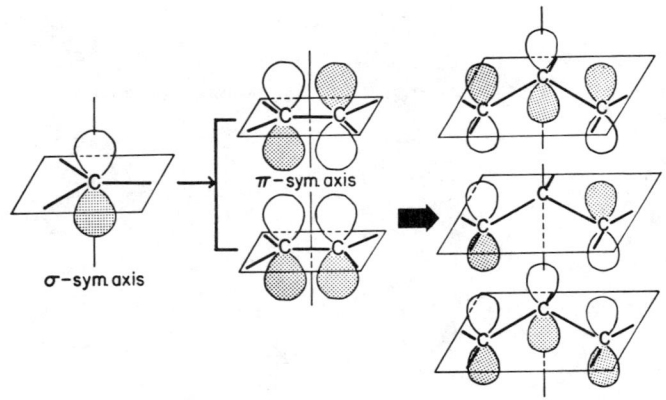

Fig. 12. The generation of molecular orbitals by the combination of carbon p orbitals.

Characteristically *transition metal atoms or ions* have orbitals with various symmetries (σ, π, or δ) to the availability of d orbitals. For example, Vaska's iridium complex $IrCl(CO)(PPh_3)_2$ is both a σ acceptor toward SO_2 and a π donor toward tetracyanoethylene (TCNE) and is simultaneously σ-accepting and π-donating toward CO (Fig. 13).

---

* "Frontier orbitals" are the highest occupied molecular orbital (HOMO) and the lowest unoccupied molecular orbital (LUMO). These two orbitals are particularly important in determining the course of a reaction.

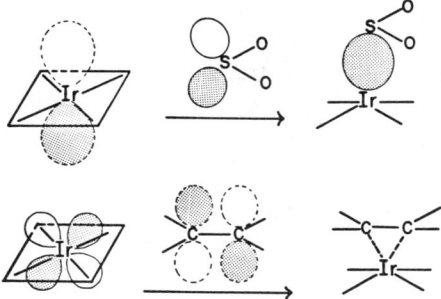

Fig. 13. A σ-acid-base interaction with a σ-acidic metal orbital and a π-acid-base interaction with a π basic metal orbital.

Some transition metal atoms are strong π acids (π acceptors) and bind firmly with organic π bases. η^6-Benzene complexes[†] of V^0, Cr^0 (Fig. 14), and Cr^I are

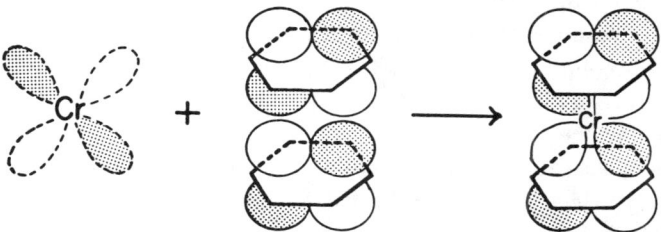

Fig. 14. The bonding in bis(arene)chromium(0) through π-acid-base interactions.

well-known examples. In the f-transition actinide series, the participation of f orbitals in bonding has been considered as a possibility. A metal f orbital in $U(C_8H_8)_2$ may be functioning as a δ acceptor as shown in Fig. 15.

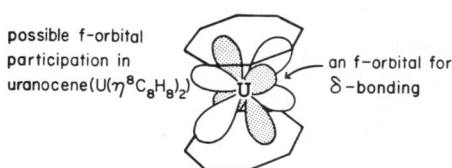

Fig. 15. A possible δ-acid-base interaction in uranocene $U(\eta^8\text{-}C_8H_8)_2$.

[†] The "η" is called "hapto" and indicates the bonding status of a multicenter ligand to a metal. The superscript indicates the number of atoms of the ligand bonded to the metal.

The number of d electrons is an important feature describing the chemistry of a d-block transition metal compound. It has become usual practice to add formal oxidation numbers to any transition metal compounds because the oxidation number will determine the *number of d electrons* and will give some indication of the *effective oxidation state* of the central metal. The number of d electrons in a metal complex is equal to: (a) in ionically bound complexes such as $[CoCl(NH_3)_5]^{2+}$, the total number of d electrons in the spin-paired metal atom minus the formal charge [e.g., +3 on the cobalt(III)], and (b) in essentially covalently bound complexes such as $Cr(C_6H_6)_2$, the number of d electrons of the spin-paired metal atom.

One can assign the number of d electrons for most transition metal complexes through reasonable approximations of the metal-ligand polarity. For example, in $Cr(CH_3)_4$ the Cr–C bond is polarized as Cr^+—C^-, and the formal oxidation state of the Cr ion is +4; therefore $Cr(CH_3)_4$ is a d^2 complex. Having assigned the number of d electrons and formal oxidation states to a particular metal complex, one can discuss the chemistry of the compound based on general principles already established in inorganic chemistry. For example, in d^{10} or high-spin d^5 complexes, the electrons are evenly dispersed in all five d orbitals and the resulting d-electron density around the metal is spherically symmetric (or totally symmetric).

Similar spherically symmetric centers can also be achieved by filling up one s and three p orbitals to give a s^2p^6 electronically closed system (e.g., Na^+, Ca^{2+}, Cl^-). In these closed-shell ions there is no steric preference or constraint caused by the orbitals for bonding with monodentate ligands. The resulting metal complex adopts the most sterically and electrostatically stabilized coordination geometry: in particular, two-coordinate, linear; three-coordinate, trigonal planar (D_{3h}); four-coordinate, tetrahedral (T_d); or six-coordinate, octahedral (O_h). Almost any coordination number (2 through 12) is possible for metal complexes with a spherically symmetric metal center. The number (and stereochemistry) of ligands is determined by a delicate balance of steric and electronic factors and is therefore variable. For example, Cu^+ (d^{10} ion) forms two-, three-, four-, and five-coordinate complexes that are very labile. This characteristic is ideally suited for any catalysis involving a metal as the activating center.

In contrast to spherical metal ions, low-spin d^6 metal ions have a strong preference for normal octahedral (O_h) structures. If there is no special steric or electronic constraints in the ligand to exclude orientation to this preferred O_h structure, it is always safe to expect regular O_h structures for any d^6 metal complex. This strong tendency is primarily due to a stable electronic closed-shell configuration, $(t_{2g})^6$ or $(d_{xy}^2 d_{yz}^2 d_{xz}^2)$, in the O_h ligand field. Figure 16 gives the

d-electron distributions schematically. This particularly stable electronic configuration renders any change disturbing it very energetically unfavorable. Dissociation of one of the ligands or association with another attacking ligand requires so much energy that the low-spin d^6-O_h complexes are generally kinetically inert. Therefore these inert complexes are not good catalysts unless activated by ligands, H or alkyl, by photochemical means, or by the addition of suitable reagents that destroy the d^6-O_h configuration.

Low-spin d^8 complexes of second- and third-row metal ions have a strong tendency to be four-coordinate with a square planar structure. These d^8 complexes, [e.g., Wilkinson's rhodium phosphine complex RhCl(PPh$_3$)$_3$] are often very important homogeneous catalysts because they have vacant coordination sites for addition of other reactants to become five-coordinate, and they can sometimes dissociate into still more reactive three-coordinate species. In the d^8 square planar complexes, one d orbital is vacant and accepts σ donation from four ligand orbitals in cooperation with vacant s and p orbitals. Figure 16 is the idealized illustration of the d^8 square planar metal core.

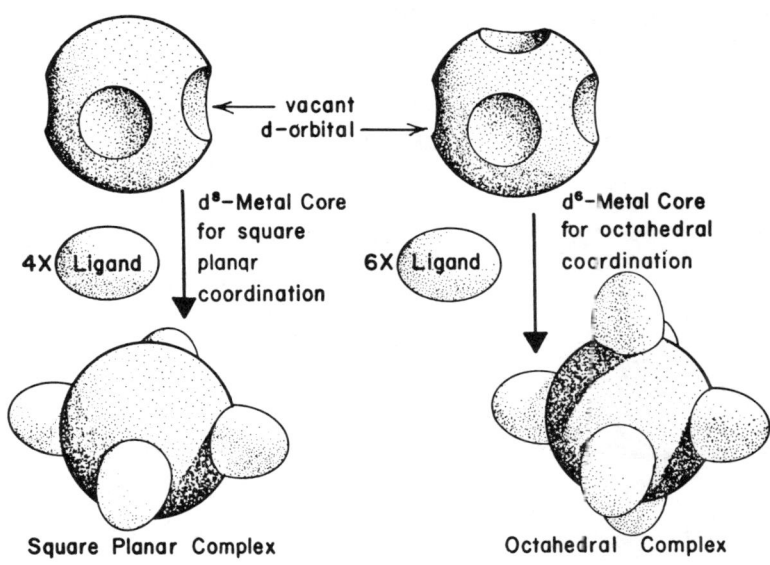

Fig. 16. A schematic illustration of d-electron distribution in low-spin d^8 and d^6 metal cores and the resulting complexes by ligation.

Modern theory of chemical reactions divides orbital interactions (see Chapter 4 also) into two distinct categories: exchange interactions and generalized donor-acceptor interactions. A typical example of the exchange interaction is the thermally allowed [$4_s + 2_s$] cycloaddition (the subscript s stands for a suprafacial reaction) or Diels-Alder reaction between butadiene and maleic anhydride. A characteristic feature of this interaction is "crossed-mutual" HOMO–LUMO interaction (Fig. 17).[15]

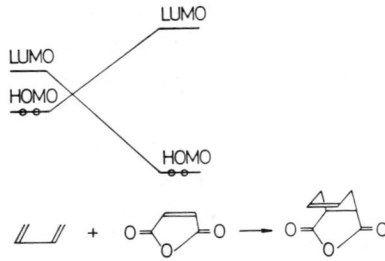

Fig. 17. A frontier orbital interaction scheme for a thermally allowed cycloaddition.

The generalized donor-acceptor (DA) interaction results from an electron transfer from the HOMO of the donor to the LUMO of the acceptor (Fig. 18).

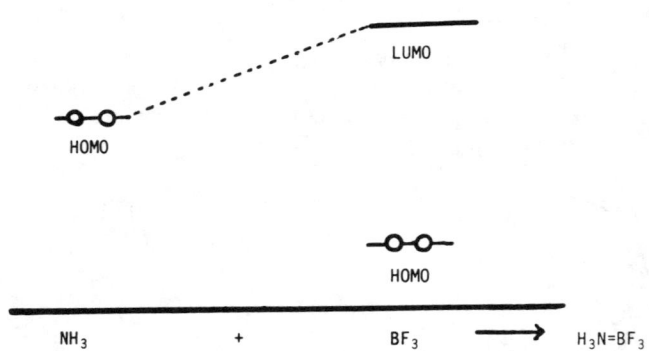

Fig. 18. An orbital interaction scheme for donor-acceptor complex formation.

The electron density is donated to the acceptor when the symmetry of the interacting orbitals matches (i.e., there is a positive or bonding interaction). The DA interaction implies that some polar character is present. In contrast, the exchange interaction means that an essentially nonpolar character is present.

However actual chemical reactions are not clear-cut, and a composite of these two types of interaction is frequently observed.

Acid-base catalysis, which is usually subdivided into specific, general, electrophilic, and nucleophilic catalyses, consists basically of DA interactions (see below). The coordination of basic ligands to a metal ion belongs to this category.

The HOMO-LUMO interaction between an essentially covalent transition metal-hydrogen bond and a C=C double bond is a key step in metal-catalyzed olefin hydrogenation (Fig. 19). η^2 Coordination of olefins to low-valent metals

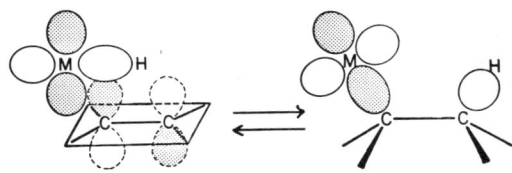

Fig. 19. Orbital interactions in a metal-alkyl complex formation by olefin insertion into the metal-hydrogen bond.

has been also regarded as a kind of the exchange interaction because two HOMO-LUMO interactions occur simultaneously but usually to different extents. As a result of the exchange interaction (Fig. 20), only a small amount of net charge transfer takes place in olefin η^2 coordination.

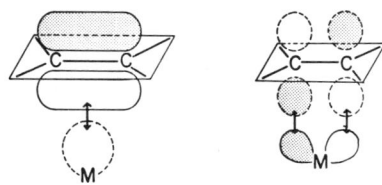

Fig. 20. Cooperative σ and π interactions in a metal-η^2-olefin bond.

Now let us consider the shape and level of metal orbitals when one ligand is removed by some means from an inert d^6-O_h complex. (This process corresponds to "photochemical activation.") Elian and Hoffmann have calculated the probable shape and level of the frontier orbitals of Cr(CO)$_5$ in a square pyramidal structure (C_{4v} symmetry).[16] The level of one of the original e_g orbitals falls and becomes a vacant a_1 orbital (LUMO) with some mixing of s and p orbitals (Fig. 21). The filled d_{xz} and d_{yz} orbitals in the t_{2g} orbitals now function

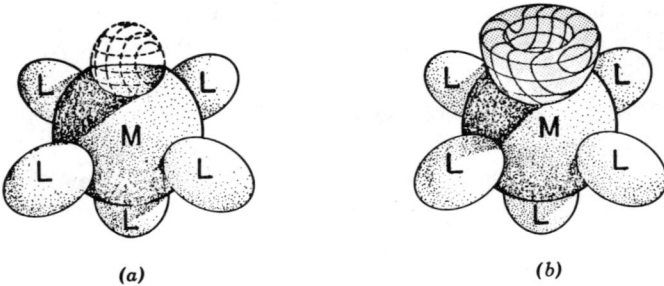

Fig. 21. The shape of frontier orbitals of a five-coordinate d^6 metal fragment in a square pyramidal structure. (*a*) Vacant orbital shown as a hole (σ symmetry). (*b*) Filled e orbital extending above in π symmetry.

as π-donating orbitals and are closer in energy to the vacant σ orbital (a_1, LUMO). The situation at the square pyramidal Cr(0) atom is ideally suited for binding a CO molecule by an exchange interaction, that is, a simultaneous σ interaction with σ-DA and a π interaction with π-DA orbitals, as shown in Fig. 22.

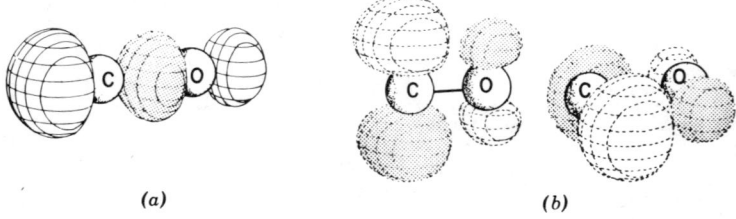

Fig. 22. The σ-donor and π-acceptor orbitals of carbon monoxide. (*a*) Filled 5 σ orbital ($E = -0.5544$). (*b*) Vacant 2 π orbitals ($E = 0.1268$). From W. L. Jorgensen and L. Salem, *The Organic Chemist's Book of Orbitals,* Academic Press, NY, 1973.

Similar shapes of frontier orbitals are probable for four-coordinate d^8–square planar complexes. It is important to note that the effective oxidation state of these square planar complexes affects the spatial extension of the frontier orbitals, thereby exerting a strong influence on interaction with an attacking reagent.

In contrast to the well-known directional property of sp^n hybrid orbitals in hydrocarbons, the metal-ligand bonds are generally much less directional. Distortion from a regular polyhedral structure, therefore, occurs readily on thermal activation. This distortion leads to conversion of one structure to another in some cases. Thus thermal distortion (as shown below) for a five-coordinate square pyramid structure smoothly leads to a trigonal bipyramid structure. In many cases these two structures are found to interconvert readily in solution.

Let us examine how the frontier orbitals are influenced by the distortion of a square planar d^8 complex. According to extensive studies by Rösch and

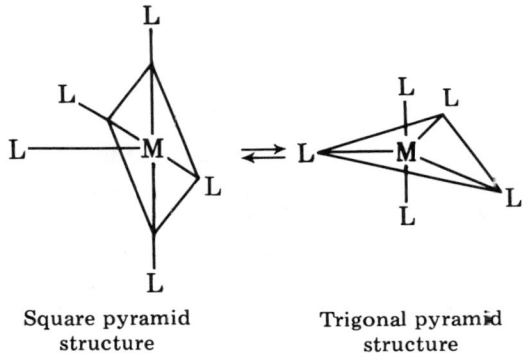

Square pyramid structure   Trigonal pyramid structure

Hoffmann, the energy level changes can be schematically shown as in Fig. 23 for a "C_{2v} distortion."[17]

An important result of the C_{2v} distortion is that the one (b_2) of the π-donating orbitals is raised in energy and is expected to work more efficiently as a π base

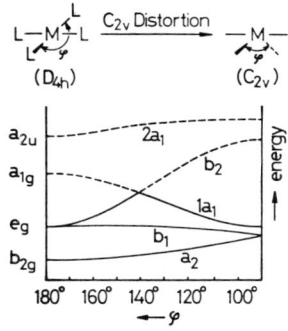

Fig. 23. The energy levels of metal orbitals in ML_4.

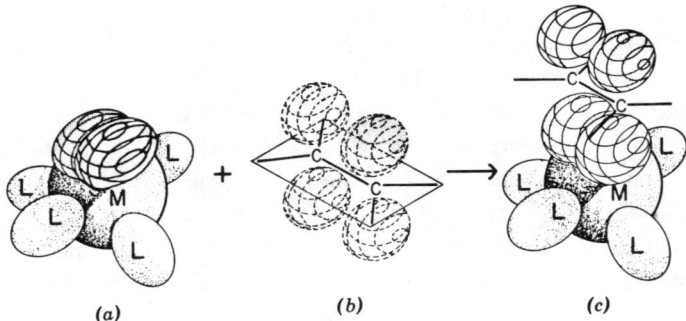

Fig. 24. The π interaction between a C_{2v}-distorted square metal complex and an olefin. (*a*) π-Donating orbital in C_{2v}-distorted ML$_4$. (*b*) π-Accepting orbital of an olefin. (*c*) η^2-Olefin complex M(η^2-Olefin)L$_4$.

in a specified direction. This situation is illustrated in Fig. 24. Thus π-acidic compounds would react more easily with the C_{2v}-distorted square planar d^8 complexes than with the undistorted ones (Fig. 25). Ligands able to undergo this C_{2v} distortion are expected to be more active for olefin or acetylene reactions. Suitable ligands can be designed by connecting the ligating atoms with organic chain(s) of appropriate length.

Since a number of labile d^8 metal complexes are utilized as homogeneous catalysts for olefin or acetylene reactions, the shape and behavior of the π-basic

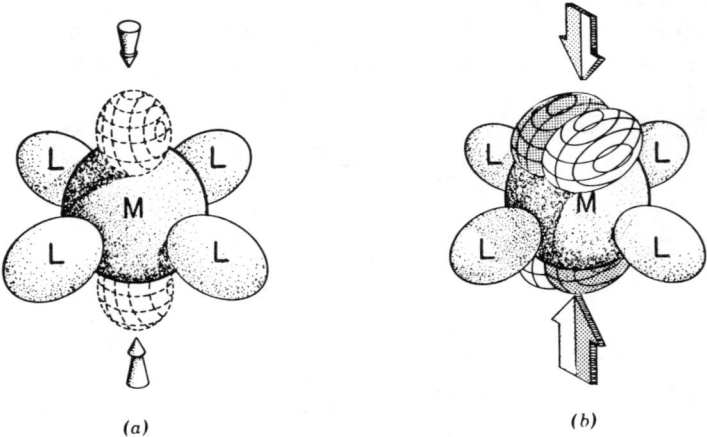

Fig. 25. Schematic illustrations of σ-accepting and π-donating metal orbitals. (*a*) σ-Accepting orbital [M^{II}L$_4$]$^{2+}$ D_{4h}. (*b*) π-Donating orbital [M^{0}L$_4$]^0 D_{4h}.

orbitals of the metal is of fundamental importance in considering mechanisms of these catalyses.

SELECTED READINGS

Kinetics

W. C. Gardner, Jr., *Rates and Mechanisms of Chemical Reactions,* Benjamin, New York, 1969.

M. Boudart, *Kinetics of Chemical Processes,* Prentice-Hall, Englewood Cliffs, N.J., 1968.

Comprehensive Kinetics, Vols. 1–14, C. H. Bamford and C. F. H. Tipper, Eds., Elsevier, Amsterdam, 1975.

B. R. James, *Homogeneous Hydrogenations,* Wiley, New York, 1973.

Stereochemistry

K. Mislow and M. Raban, in *Topics in Stereochemistry,* N. L. Allinger and E. L. Eliel, Eds., Vol. 1, Springer, Berlin, 1967, p. 1.

E. L. Eliel, *Stereochemistry of Carbon Compounds,* McGraw-Hill, New York, 1966.

C. J. Hawkins, *Absolute Configurations of Metal Complexes,* Wiley-Interscience, New York, 1971.

Y. Izumi and A. Tai, *Stereo-Differentiating Reactions,* Academic Press, New York, 1976.

Y. Izumi, *Agnew. Chem., Int. Ed., 10,* 871 (1971).

Reactivity Theory

R. G. Pearson, *Symmetry Rules for Chemical Reactions,* Wiley-Interscience, New York, 1976.

G. Klopman, in *Chemical Reactivity and Reaction Paths,* G. Klopman, Ed., Wiley-Interscience, New York, 1974, pp. 55.

L. Salem and W. L. Jorgensen, *An Organic Chemist's Book of Orbitals,* Academic Press, New York, 1973.

References for Chapter 3

1. A. Nakamura, K. Doi, K. Tatsumi, and S. Otsuka, *J. Mol. Catal., 1,* 417 (1976).
2. C. O'Connor and G. Wilkinson, *J. Chem. Soc. A, 1968,* 2665.
3. J. Halpern, in *Organo Transition Chemistry,* Y. Ishii and M. Tsutsui, Eds., Plenum Press, New York, 1975, pp. 109.
4. J. Hjortkjaer, *Homogeneous Catalysis-*II, D. Forster and J. F. Roth, Eds., American Chemical Society *Advances in Chemistry* Series, Vol. 132, ACS, Washington, D.C., 1974, pp. 133; D. E. Rudd, D. G. Holah, A. N. Hughes, and B. C. Hui, *Can. J. Chem., 52,* 775 (1974).

5. F. H. Jardine, J. A. Osborn, and G. Wilkinson, *J. Chem. Soc., A, 1967,* 1574; B. R. James, *Homogeneous Hydrogenation,* Wiley, New York, 1973.
6. A. White, P. Handler, and E. L. Smith, *Principles of Biochemistry,* 5th ed., McGraw-Hill, New York, 1973, pp. 234.
7. C. A. Tolman and W. D. Seidel, *J. Am. Chem. Soc., 96,* 2774 (1974).
8. R. Good and A. E. Merbach, *Inorg. Chem., 14,* 1030 (1975).
9. S. Takeuchi, Y. Ohgo, and J. Yoshimura, *Bull. Chem. Soc. Japan, 47,* 463 (1974).
10. H. Wakamatsu, *J. Chem. Soc. Japan, 85,* 227 (1964).
11. H. Aoi, M. Ishimori, and T. Tsuruta, *J. Organometal. Chem., 85,* 241 (1975).
12. J. W. Teipel, G. M. Hass, and R. L. Hill, *J. Biol. Chem., 243,* 5684 (1968).
13. G. Gokel, P. Hoffmann, H. Kleimann, H. Klusacek, D. Marquarding, and I. Ugi, *Tetrahedron Lett., 1970,* 1771.
14. E. L. Eliel, *Stereochemistry of Carbon Compounds,* McGraw-Hill, New York, 1962, Chapters 4, 6, and 8.
15. I. Fleming, *Frontier Orbitals and Organic Chemical Reactions,* Wiley, New York, 1976.
16. M. Elian and R. Hoffmann, *Inorg. Chem., 14,* 1058 (1975).
17. N. Rösch and R. Hoffmann, *Inorg. Chem., 13,* 2656 (1974).

4

ELEMENTARY PROCESSES

1. GENERAL INTERACTIONS

1.1. Electrophilic and Nucleophilic Interactions

In many homogeneous catalysis systems proton transfer occurs between reactants and catalysts. For example, protonation of a carbonyl group results in higher reactivity at the carbonyl carbon toward nucleophilic attack. When conditions are suitable, the protonated ketone may react with the nucleophile.

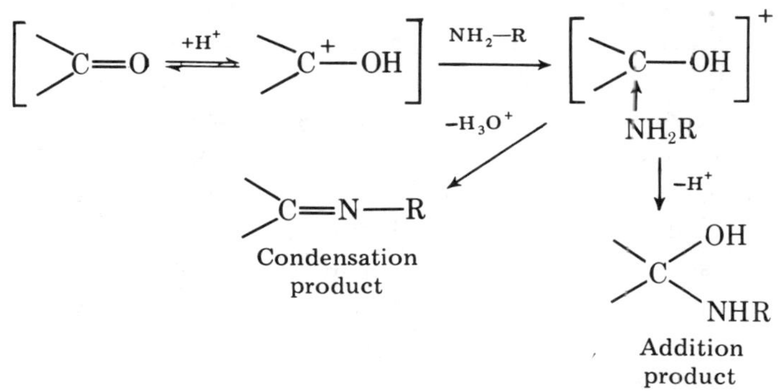

$$\left[\diagup\!\!\!\!\diagdown\!\!C\!=\!O \underset{}{\overset{+H^+}{\rightleftarrows}} \diagup\!\!\!\!\diagdown\!\!\overset{+}{C}\!-\!OH \right] \xrightarrow{NH_2-R} \left[\diagup\!\!\!\!\diagdown\!\!C\!-\!OH \atop NH_2R \right]^+$$

with branches leading to:

- $\diagup\!\!\!\!\diagdown C\!=\!N\!-\!R$ (Condensation product), via $-H_3O^+$
- $\diagup\!\!\!\!\diagdown C\!\!\diagup^{OH}_{NHR}$ (Addition product), via $-H^+$

Scheme 1

Table 2. Brown's Selectivity Relationship

Electrophile		Relative Rate $k_{toluene}/k_{benzene}$	Product Distribution (%)	
			Meta	Para
Br^+	(bromination)	605	0.3	66.8
$PhCO^+$	(benzoylation)	110	1.5	89.3
NO_2^+	(nitration)	23	2.8	33.9
Hg^{2+}	(mercuration)	7.9	9.5	69.5
$\mathrm{CH_3{>}\overset{+}{C}{-}H\atop CH_3}$	(isopropylation)	1.8	25.9	46.2

Electrophiles such as Mg^{2+} or Zn^{2+} are particularly effective in catalyzing the reactions above. Reactions of this type are called acid-catalyzed or electrophilically catalyzed reactions. The proton is the simplest electrophilic catalyst, but cannot work by itself because it is always associated with a nucleophilic partner, called a conjugate base. Therefore not only protons but also the conjugate bases must be taken into account in interpreting acid-catalyzed reactions.

Since there is great variety among the acid-base interactions occurring between hundreds of known catalysts and substrates, the descriptions here are confined to a few simple and fundamental cases. Modern molecular orbital (MO) theory interprets these interactions via a simple "generalized donor-acceptor" orbital interaction scheme.

Let us begin our survey with the proton acidity of various acids (general formula, HX). A quantitative measure of the extent of acid-base interactions in compounds having dissociable protons is the pK_a value. The pK_a value[1] ranges widely from $-10(HClO_4)$ to $44(Me_2CH{-}H)$. Selected pK_a values are: HCl, -7; H_3O^+, -1.74; HF, 3.17; RCO_2H, 4 to 5; $[WH_3(Cp)_2]^+$, 5.4; H_2CO_3, 6.35; NH_4^+, 9.24; RSH, 12; $CH_2(CN)_2$, 12; H_2O, 15.7; $PhC{\equiv}CH$, 18.5. There are many other kinds of electrophilic center, for example, the carbon of a carbonyl group, and the nitrogen of a nitronium ion (NO_2^+). One measure of quantitative assessment of the strength of donor-acceptor interactions involving these electrophiles with positive carbon or nitrogen can be obtained by examining competitive Friedel-Crafts reactions, which proceed by the intermediacy of these electrophilic species. Thus the molar ratio of the product of a Friedel-Crafts reaction (see below) in a mixture of benzene and toluene reflects the electrophilic strength of the attacking electrophile toward an sp^2 hybridized carbon (aromatic nucleus). Increased selectivity (high k_{tol}/k_{ben} values and high para/meta ratio, see Table 2) indicates weaker electrophilicity.[2]

General Interactions

Electrophile X (e.g., RCO^+ or NO_2^+)

[Reaction schemes showing electrophilic aromatic substitution of benzene and toluene giving ortho, meta, and para products]

In these Friedel-Crafts reactions the efficiency of the Lewis acid catalyst is exhibited by the relative strength of its electrophilic interaction with the substrate (e.g., alkyl halides).

$$2MX_n \rightleftharpoons MX_{n-1}^+ + MX_{n+1}^-$$

$$MX_{n-1}^+ + RX \rightleftharpoons R^+ + MX_n$$

Some common Lewis acid catalysts, in decreasing order of efficiency, are $AlBr_3$, $AlCl_3$, $GaCl_3$, $FeCl_3$, $SbCl_5$, $ZrCl_4$, BCl_3, BF_3. This order is only approximate: it also depends on substrates and conditions.[3]

Among the dipositive transition metal ions (M^{2+}) of the first row, the ability to coordinate with various polar donor molecules (e.g., NH_3 or pyridine) in water provides a sequence of acceptor ability for σ-donor ligands. This sequence is known as the Irving-Williams series, in which the relative order of stability of NH_3 complexes is $Zn^{2+} < Cu^{2+} > Ni^{2+} > Co^{2+} > Fe^{2+} > Mn^{2+}$.[4] This series, however, is applicable only to limited cases. There is no corresponding series for the second- or third-row transition metal ions, because aquated dipositive metal cations are not always available in aqueous solutions.

There are a number of catalytically active second- or third-row transition metal complexes that have little Lewis acid (or electrophilic) character at the metal. The weak electrophilic interaction with a hard base (e.g., py) has been measured utilizing the kinetics of a ligand substitution process.

[Square planar complex diagram: L–M–X with Y and L ligands + Nu → [L–M–Nu with Y and L ligands]$^+$ + X^-]

Among the substitution reactions of square planar d^8 noble metal complexes, the best suited reaction has been nucleophilic displacement of halide ligands in aqueous or alcoholic solution.[5] Platinum(II) complexes have been generally favored for such kinetic studies because of their convenient range of reaction rates. The rates (k_{obs}) observed in the reaction shown below are apparently first order in each complex, as long as the nucleophile is present in excess. The relationship can be summarized by the following equation:

$$k_{obs} = k_1 + k_2[N], \qquad [N]: \text{concentration of the nucleophile}$$

where k_1, the first-order rate constant, refers to the rate constant for the nucleophile-independent path, and k_2, the second-order rate constant, can be derived from the slope of a plot of k_{obs} versus [N]. Two reaction paths are possible, one involving the solvent and the other involving the nucleophile in the rate-determining step. These two rate constants can span a considerable range, depending on the identity of metal, its oxidation state, its auxiliary ligands, and, of course, the nucleophilicity of the nucleophile. If the nucleophile is pyridine, then k_1 and k_2 for chloride substitution in *trans*-PtCl(L)(PEt$_3$)$_2$ are as follows:

$$\left[\begin{array}{c} \text{PEt}_3 \\ | \\ \text{L—Pt—Cl} \\ | \\ \text{PEt}_3 \end{array}\right] + \text{py} \longrightarrow \left[\begin{array}{c} \text{PEt}_3 \\ | \\ \text{L—Pt—py} \\ | \\ \text{PEt}_3 \end{array}\right] + \text{Cl}^-$$

L	k_1 (sec^{-1})	k_2 (l mol^{-1} sec^{-1})
PEt$_3$	1.7×10^{-2}	3.8
H	1.8×10^{-2}	4.2
CH$_3$	1.7×10^{-4}	6.7×10^{-2}
Cl	1.0×10^{-6}	4.0×10^{-4}

Reference 6.

The strong accelerating effect of a trans ligand on the leaving Cl$^-$ is called the trans effect. Hydrido, σ-alkyl, η^2-olefin, and phosphines are among ligands with a strong trans effect. The identity of the leaving ligand (X$^-$) also affects the rates, as the observed rates in the following reaction indicate:

$$\left[\begin{array}{c} \text{NH}_2 \\ | \\ \text{HN——Pt—X} \\ | \\ \text{NH}_2 \end{array}\right]^+ + \text{py} \longrightarrow \left[\begin{array}{c} \text{NH}_2 \\ | \\ \text{HN——Pt—py} \\ | \\ \text{NH}_2 \end{array}\right]^{2+} + \text{X}^-$$

General Interactions

	$k_{obs} \times 10^6$ (sec^{-1})		
X = NO$_3^-$	Immeasurably fast		
X = Cl$^-$	35	X = Br$^-$	23
X = SCN$^-$	0.30	X = CN$^-$	0.017

The larger observed rate constants indicate more effective electrophilicity toward pyridine. Auxiliary ligands can also exert remarkable steric effects. Thus as the following rates show, increasing the steric bulk of the σ-aryl ligand in cis- and trans-PtCl(aryl)(PEt$_3$)$_2$ greatly reduces the rate of substitution of Cl$^-$ with pyridine. These results also support the associative nature of the reaction because the "effective" electrophilicity at the metal center is decreased by steric hindrance.[8]

k_{obs} (sec^{-1}) at 25°C: 3.4 × 10^{-6} 1.7 × 10^{-5} 1.2 × 10^{-4}

k_{obs} (sec^{-1}): 1.0 × 10^{-6} (at 25°)   2.0 × 10^{-4} (at 0°)   8.0 × 10^{-2} (at 0°)

The nature of nucleophiles as reflected by the rate of various substitution reactions has been extensively studied by Edwards and Pearson.[9a] The classification of acids and bases as "soft" and "hard" has been the most important qualitative result of their study. *Soft* nucleophiles (S^{2-}, I$^-$, etc.) are easily polarized and are most effective toward *soft* (easily polarized) substrates (e.g., CH$_3$Hg$^+$); similarly *hard* nucleophiles (e.g., OH$^-$, NH$_3$) are difficult to polarize and are most effective toward hard substrates (e.g., H$^+$ or

$$R-\underset{\underset{OR}{|}}{C}=O).$$

For a series of organic substitution reactions, Swain and Scott have also reported a nucleophilicity scale toward carbon for various nucleophiles reacting with methyl bromide (a relatively soft substrate) (Table 3). A larger n implies a "softer" nucleophile.[9b]

Another nucleophilicity scale for substitution in square planar Pt(II) complexes has been advanced utilizing kinetic data such as those described above. A nucleophilic reactivity constant n_{Pt}^0 (defined as shown below), has been devised.[10]

$$\log\left(\frac{k_Y}{k_S}\right) = n_{Pt}^0$$

The k_Y and k_S values refer to rate constants for substitution of Cl^- by a nucleophile and solvent in $trans$-$PtCl_2(py)_2$ in MeOH at 30°. Typical n_{Pt}^0 values are listed in Table 4, together with comparable values of $n_{CH_3I}^0$, which are for the analogous reaction with the same nucleophiles toward CH_3I. Some anionic metal chelates have a very high nucleophilicity. For example, natural vitamin B_{12s} and cobinamide both have $n_{CH_3I}^0$ values of 14.4, which means that they react some 10^{14} times faster than OH^-. Synthetic model compounds [cobaloxime-

Table 3. Swain-Scott Nucleophilic Reactivity Parameter n

Nucleophile	n
H_2O	0
AcO^-	2.72
Cl^-	3.04
Br^-	3.89
OH^-	4.20
I^-	5.04

$$\log k/k_0 = sn$$

where k and k_0 = rate constants for reaction in question and for a standard, respectively

s = substrate constant; in this case for CH_3Br, $s = 1$

General Interactions

Table 4. Examples of Nucleophilic Reaction Constants and pK_a Values[12]

Nucleophile	n^0_{Pt}	$n^0_{CH_3I}$	pK_a
AcO$^-$	<2.4	4.3	4.75
CH$_3$O$^-$	<2.4	6.29	15.8
NH$_3$	3.06	5.50	9.25
Me$_2$S	4.73	5.54	−5.3
CN$^-$	7.0	6.7	9.1
Ph$_3$P	8.79	7.0	2.61
Et$_3$P	8.85	8.72	8.86
PhS$^-$	7.17	9.92	6.52
Ph$_3$Sn$^-$	—	11.5	—

(py)]$^-$ and Co^I (salen) have similar values, 13.8 and 14.6, respectively, and these compounds are called "supernucleophiles."[11]

Edwards has utilized an extensive collection of kinetic and thermodynamic data from many organic as well as inorganic reactions to derive the following equations, known as Edwards's equation or the oxibase scale (see Table 5).

The E_n values are electron donor constants that are measures of *softness* of the nucleophiles.

Edwards's Equation (oxibase scale):

$$\log (k/k_0) = \alpha P + \beta H; \log (k/k_0) = \alpha E_n + \beta H$$

Table 5. Electron Donor Constants E_n [13]

Nucleophile	pK_a[a]	E_n	$pK_{CH_3Hg^+}$[b]	$(pK_{CH_3Hg^+})-pK_a$
S^{2-}	12.9	3.08	21.3	+8.4
CN$^-$	9.1	2.79	14.0	+4.9
I$^-$	(−10)	2.06	8.66	+18.7
OH$^-$	15.7	1.65	9.42	−6.3
Br$^-$	(−7)	1.51	6.67	+13.7
NH$_3$	9.5	1.36	7.65	−1.8
Cl$^-$	(−4)	1.24	5.30	+9.3
py	5.2	1.20	4.80	+0.4
H$_2$O	−1.7	0.00	−1.74	0.00
Et$_3$P	8.86	—	15.00	+6.1

[a] For conjugate acid of base, in H$_2$O at 25°; numbers in parentheses are estimates.
[b] Dissociation constant for the process CH$_3$HgL$^+$ → CH$_3$Hg$^+$ + L.

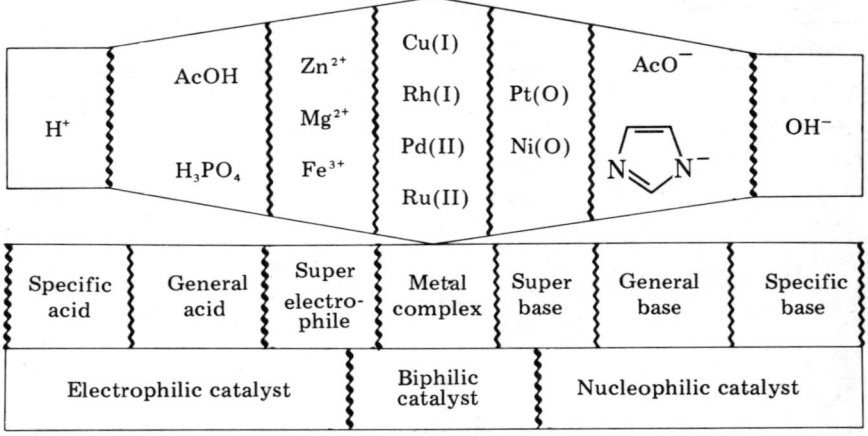

Scheme 2. A general interrelation scheme of acid-base catalysts.

where α, β = reaction constants
P = polarizability factor
H = proton basicity factor defined by $H = 1.74 + pK_a$
E_n = redox factor defined by $E_n = E^0 + 2.6$
E° = standard oxidation potential for the process $2X^- \rightarrow X_2 + 2e^-$

and k_0 refers to water; k refers to various nucleophiles.

Large values of P or E_n mean the nucleophile is *soft*. Also, equilibrium constants with a typical soft acceptor such as CH_3Hg^+ were measured and processed to yield $pK_{CH_3Hg^+}$ values that measure the *softness* of the nucleophile in question. It should be remembered that these values are obtained only in aqueous solutions where many catalytically important organometallic reactions do not occur.

To make above concepts clear, various acid-base catalysts are interrelated in Scheme 2.

1.2. Electron Donor-Acceptor Interactions

Usually a donor-acceptor (DA) interaction occurs when there is a positive electronic interaction between the HOMO of a donor and the LUMO of an acceptor. The extent of the DA interaction is essentially determined by the energy level difference between these relevant orbitals and by the effective overlap between them. An electron donor-acceptor interaction occurs (Fig. 26) when the overlap is not enough or not energetically favorable, and the energy difference

General Interactions

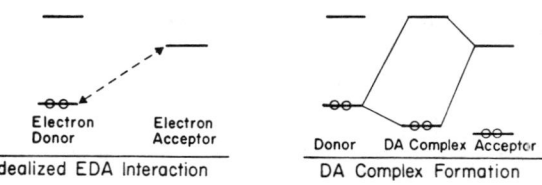

Fig. 26. An illustration of the energy levels in an idealized electron donor-acceptor (EDA) interaction and in a DA complex formation.

is sufficient to cause energetic stabilization by transferring some electron density from the HOMO to the LUMO. This partial electron transfer is called charge transfer.

If the difference between the HOMO and LUMO energy levels is small enough to overcome the total energy required to remove one electron from the donor and to add one electron to the acceptor, *one electron transfer* occurs. When electron transfer occurs between organic compounds, a radical cation and a radical anion result (Fig. 27). These radical species are generally very reactive. Some homogeneous catalysts can activate substrates through electron transfer to particular reactants.

Electron transfer between two different metal atoms or ions is one of the most important reactions in transition metal chemistry. The rates of the electron transfer between a large variety of aquated or ligated metal ions have been measured and have been found to be strongly dependent on the identity of the metal species and on the formal oxidation states. The examples presented in Table 6 illustrate this situation.

The fastest rate given in Table 6 approaches the rate of a diffusion-controlled bimolecular process (i.e., $10^9 \sim 10^{10}$ l mol^{-1} sec^{-1}). The rate constant for the reaction of a hydrated electron with a transition metal ion is mostly within this diffusion-controlled range.

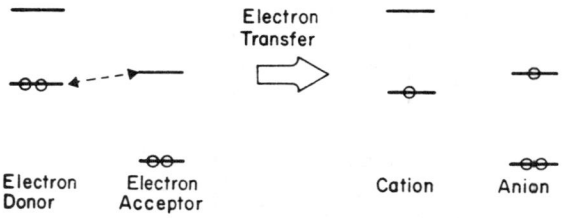

Fig. 27. An ET interaction resulting in formation of a pair consisting of a cation radical and an anion radical.

Table 6. Rates of Redox Reactions at 25° in Water[15]

Reductant	Configuration	Oxidant	Configuration	Second-Order Rate Constant k_1 (l mol^{-1} sec^{-1})
Fe^{2+}	d^6	Mn^{3+}	d^4	1.7×10^4
$[Fe^{II}(CN)_6]^{2+}$	$(t_{2g})^6$	$[Fe^{III}(phen)_3]^{3+}$	$(t_{2g})^5$	10^8
$[Co^{II}(NH_3)_n]^{2+}$	$(t_{2g})^5(e_g)^2$	$[Co^{III}(NH_3)_6]^{3+}$	$(t_{2g})^6$	10^{-9}
$Cu^{I}Cl_2^-$	d^{10}	$Cu^{II}Cl_4^{2-}$	d^9	5×10^7
Cyt-$c^{II\,a}$	$(t_{2g})^6$	Cyt-$c^{III\,a}$	$(t_{2g})^5$	3×10^3
Fe^{2+}	$(t_{2g})^4(e_g)^2$	$[Fe^{III}(edta)]^-$	$(t_{2g})^3(e_g)^2$	4×10^{-4}

^{a} Cyt-c^{II} = cytochrome c Fe^{II}, Cyt-c^{III} = cytochrome c Fe^{III}.

$$[Fe(H_2O)_6]^{2+} \xrightarrow{e^-} [Fe(H_2O)_6]^+ \quad k = 3.5 \times 10^8$$

$$[Co(H_2O)_6]^{2+} \xrightarrow{e^-} [Co(H_2O)_6]^+ \quad k = 1.2 \times 10^{10}$$

$$[Cu(H_2O)_6]^{2+} \xrightarrow{e^-} [Cu(H_2O)_6]^+ \quad k = 3.0 \times 10^{10}$$

Electron transfer between different oxidation states of the same metal species can be a very slow process. For example, the rate (at 25°) between aquated Cr^{2+} and Cr^{3+} is less than 2×10^{-5} l mol^{-1} sec^{-1}, and the rate between aquated Fe^{2+} and Fe^{3+} is 4.0 l mol^{-1} sec^{-1}.[16] These slow rates have been explained on the basis of the Franck-Condon principle, which states that electron movement itself is much faster than the thermal vibrations of atoms (such movement is called "a vertical process").[17] When the metal-ligand bond length varies with the oxidation state, some energetic excitation (activation energy $\Delta H^{\ddagger}$) is necessary for electron transfer to take place. This energy amounts to 10.8 kcal for the aquated Fe^{2+}/Fe^{3+} reaction. This situation is illustrated in Fig. 28, which indicates that acti-

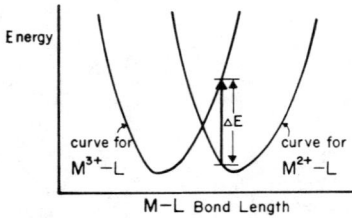

Fig. 28. An illustrative example of the energy barrier involved in electron transfer between $M^{2+}L$ and $M^{3+}L$ with different M—L bond lengths.

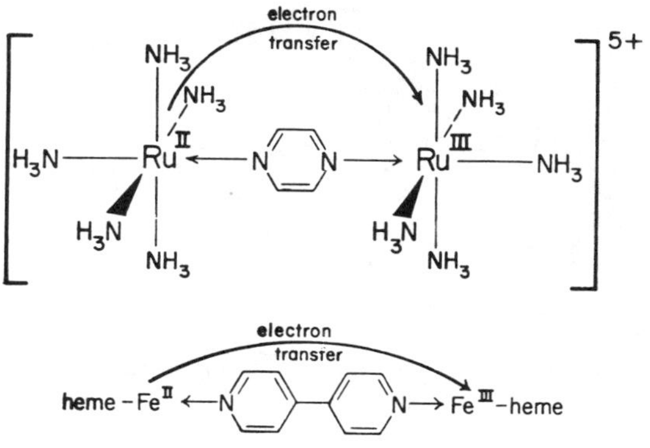

$$\left[H_3N-\underset{\underset{NH_3}{\underset{|}{H_3N}}}{\overset{\overset{NH_3}{\overset{|}{NH_3}}}{\overset{electron\ transfer}{Ru^{II}}}}\leftarrow N\underset{}{\bigcirc} N\rightarrow \underset{\underset{NH_3}{\underset{|}{H_3N}}}{\overset{\overset{NH_3}{\overset{|}{NH_3}}}{Ru^{III}}}-NH_3\right]^{5+}$$

$$heme-Fe^{II} \xleftarrow{} N\bigcirc\bigcirc N \xrightarrow{electron\ transfer} Fe^{III}-heme$$

Scheme 3

vation is required to transfer an electron from the M^{2+}—L bond length energy curve to the M^{3+}—L curve by a vertical process. As can be seen, no activation energy is anticipated when the M—L bond lengths are exactly the same before and after electron transfer between the same metal ions of different oxidation states. Electron transfer between metal clusters ($[M_nL_m]^{q+}$) seems to belong to this category of low activation energy. Since several metals participate in one complex ion, the effect of a one-electron transfer does not cause any significant change in the relevant M—L and M—M bond lengths. Metal clusters are thus expected to be good catalysts in electron transfer catalysis.[18]

The intramolecular redox reaction is remarkably enhanced by a bridging conjugated organic group. Thus the frequency of intramolecular electron transfer in a ruthenium mixed-oxidation-state complex (Scheme 3) is very large (5×10^9 sec^{-1}). Similarly, when Fe^{III} heme and Fe^{II} heme were connected by 4,4'-dipyridyl (Scheme 3), the frequency of electron transfer was estimated to be $>6 \times 10^9$ sec^{-1}.[19]

Biological electron transfer pathways involving cytochrome c^{III} have been studied by chemical methods, and the process seems to involve direct electron transfer to the Fe^{III} center through a suitable bridging group (Fig. 29) or along the exposed edge of the porphyrin ligand.[20]

Electron transfer between transition metal ions or complexes and organic molecules has an important bearing on transition-metal catalyzed organic reactions. The air oxidation of olefins catalyzed by Co(II) or Mn(II) alkanoate involves various types of electron transfer among the metal, the olefin, and oxygen.[21] The most important step here is electron transfer from the olefin to the reactive Co(III) species, which has itself been formed by an electron transfer from Co(II) to oxygen.

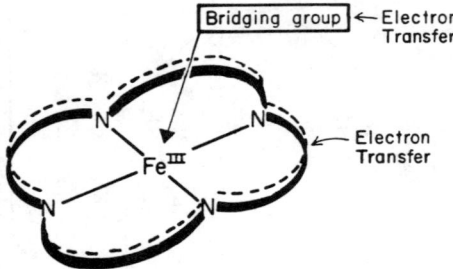

Fig. 29. Two possible paths for electron transfer into an Fe^{III}-containing site of cytochrome c^{III}.

$$RCH=CH_2 + [Co^{III}(O_2CR)_2]^+ \rightarrow R-\overset{+}{C}H-\dot{C}H_2 + Co^{II}(O_2CR)_2$$

Electron transfer reactions from an organic free radical ($R\cdot$) of the growing polymer chain

$$(CH_2-\underset{\underset{CONH_2}{|}}{CH})_n-CH_2-\underset{\underset{CONH_2}{|}}{\dot{C}H}$$

to some transition metal ions have been kinetically measured. The second-order rate constants are large and have low activation energies.[22]

	k (l mol^{-1} sec^{-1})	E_a (kcal)
$R\cdot + Fe(H_2O)_6^{3+}$	2.8×10^3	2.3
$R\cdot + Fe(phen)_3^{3+}$	3.1×10^5	4.3
$R\cdot + Cu(H_2O)_4^{2+}$	1.2×10^3	5.4

These results explain the effective inhibition (negative catalysis) of radical chain polymerization of vinyl monomers by the transition metal ions.

1.3. Radical Interactions

Free radicals (or radicallike species) are frequently involved in catalysis and, a brief account of interactions of radicals with nonradicals, metal atoms or ions, metal complexes, or other radicals is appropriate here.

The radical interactions are classified as follows:

1. An attacking radical can abstract a neutral atom (e.g., a hydrogen or a halogen from an organic structure) and produce in turn a different radical.

Table 7. Dissociation Energies[23]

R—H	Dissociation Energy (kcal/mol)
HC≡C—H	121
CH_3—H	102
Ph—H	102
Et—H	98
iPr—H	94
t-Bu—H	90
$PhCH_2$—H	78
Allyl-H	77

The relative stability of hydrocarbon radicals is

$$R· + H—R' \rightarrow R—H + R'·$$
$$R· + X—R' \rightarrow R—X + R'·$$

defined by the dissociation energies of R—H bonds. Table 7 gives some examples of the dissociation energies.

2. When two radicals combine, if they are of the same type, the process is called radical recombination, and if the radicals are of different types, the process is called radical heterocoupling:

$$R· + R· \rightarrow R—R \quad \text{Radical homocoupling, recombination, or dimerization}$$

$$R· + R'· \rightarrow R—R' \quad \text{Radical heterocoupling}$$

3. A radical can disproportionate to two nonradicals when radical homocoupling is hindered by the steric bulk of the radical. Thus the t-butyl radical $(CH_3)_3C·$ will disproportionate to isobutane and isobutene.

4. A radical can add to an unsaturated bond or to a metal atom or ligand atom.[21]

$$R· + \hspace{-2pt}\diagdown\hspace{-6pt}_{\diagup}C{=}C\hspace{-6pt}\diagup^{\diagdown} \rightarrow \hspace{-2pt}\diagdown\hspace{-6pt}^{R}C{-}C·\hspace{-6pt}\diagup^{\diagdown}$$

$$R· + ML_n \rightarrow R—ML_{n-1} + L$$
$$R· + CuCl_2 \rightarrow R—Cl + CuCl$$

Some molecules have exceptionally high reactivity toward free radicals, a well-known example being the nitroso compounds. Nitroso-t-butane in particular has been widely utilized for trapping free radicals, yielding a stable free radical that can be easily detected via esr spectroscopy.[24]

$$R\cdot + t\text{-Bu}-N=O \longrightarrow \underset{R}{\overset{t\text{-Bu}}{>}}N-O\cdot$$

5. A radical can remove one electron from an electron-donating reactant or lose one by transfer to an electron-accepting molecule. The radical can then become an anion (R^-) or a cation (R^+). This type of electron transfer frequently occurs between organic and inorganic radicals because the two differ widely in electronegativity. Some examples can be found in electron transfer redox catalysis with some inorganic reagents.[25]

$$OH\cdot + \text{[naphthalene]} \longrightarrow OH^- + \text{[naphthalene]}^{\cdot+}$$

$$O_2 + Co^{2+} + NH_3 \longrightarrow [(NH_3)_5Co^{III}-(O_2^-)-Co^{III}(NH_3)_5]^{5+}$$

$$R-CH_2\cdot + Co^{3+} \longrightarrow R-\overset{+}{C}H_2 + Co^{2+}$$

There are a great variety of radicals differing in structure and reactivity. The simplest, the hydrogen (H·), has been shown to be very reactive and occurs as an intermediate species in some gas phase reactions. The methyl radical as generated by the homolytic decomposition of diacetylperoxide [$(CH_3CO)_2O_2$]

$$(CH_3CO)_2O_2 \rightarrow 2CH_3CO_2\cdot \rightarrow 2CH_3\cdot + 2CO_2$$

can generally abstract hydrogen from almost all organic compounds, or can add to unsaturated bonds or to metal atoms. These highly reactive free radicals are *not* involved in homogeneous catalysis. Catalysis takes place utilizing radical species that are stabilized radicals, cation radicals, or anion radicals.

The reactivity of complex organic radicals when compared to that of simple saturated radicals like the methyl radical is greatly reduced by the introduction of sterically and electronically moderating substituents. Thus delocalization of the radical density at $RCH_2\cdot$ by substituents like phenyl or vinyl group(s) remarkably stabilizes the radical. Construction of special highly delocalized organic skeletons together with suitable steric protection of the radical center(s) ultimately makes organic radicals stable to most reagents, particularly oxygen.

General Interactions

The following examples of stable organic radicals have been called π radicals because the unpaired electron is delocalized in a $p\pi$ orbital:

Galvinoxyl   1,3,5-Triphenylverdazyl   di-tert-Butylnitroxide

These stable free radicals and similar derivatives have been used to detect, by radical coupling, more reactive free radicals formed in catalytic reactions.

With decreasing radical reactivity, abstraction and addition become less favorable. Stable radicals do not normally abstract hydrogen from organic reagents and do not enter into radical coupling reactions. Metal-centered radicals such as $\cdot Si(CH_2SiMe_3)_3$ have been also regarded as stable radicals.[27] Paramagnetic transition metal ions have been considered as stable metal-centered radicals. By suitable choice of ligand or anion, these paramagnetic ions can generate reactive free radicals through one-electron transfer.

$$Mn^{3+} + R-CH=CH_2 \rightarrow Mn^{2+} + R-\overset{+}{C}H-\dot{C}H_2$$

In particular, a cobalt(III) ion surrounded by ligands containing oxygen as the coordinating atom [e.g., $Co(OAc)_3$] is "paramagnetic" and has a strong tendency to generate organic radical cations by electron transfer.[28]

There have been some examples of paramagnetic metal compounds that catalyze reactions only by virtue of an unpaired electron spin. Ortho-para conversion of hydrogen is one well-known example involving the exchange of the nuclear spin of the hydrogen nucleus. Since a smaller amount of energy is associated with nuclear spins,[29] one of the hydrogen nuclear spins is readily exchanged by the influence of the unpaired *electron* (Scheme 4) on the metal.

Oxygenation of certain dienes by the paramagnetic dioxygen molecule has

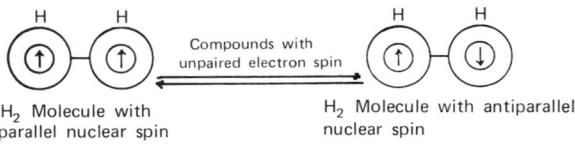

Scheme 4

been shown to be strongly affected by the presence of a catalytic amount of paramagnetic metal compounds.[30] It is suggested that the paramagnetic complex catalyzes a singlet-triplet conversion of the excited dioxygen.

The cis-trans isomerization of azobenzene has also been catalyzed by some paramagnetic metal complexes. Delocalization of the electron spin density of the metals to the η^1-coordinated azobenzene seems to lower the barrier for the geometric isomerization.[31]

2. ELEMENTARY REACTIONS IN TRANSITION METAL CHEMISTRY

2.1. Coordination and Dissociation of Ligands

Homogeneous catalysis with transition metal complexes generally involves many different chemical equilibria of coordination and dissociation througout the whole catalytic cycle. Proper understanding of these equilibria is a prerequisite for a study of homogeneous catalysis. In some labile low-valent d^8 or d^{10} metal complexes, ligand dissociation equilibria have already been extensively investigated because of their importance in catalyzing various reactions. For example, the following coordination-dissociation preequilibria have been proposed for the well-known homogeneous hydrogenation catalyst system involving $RhCl(PPh_3)_3$-H_2-olefin.[32]

1. $RhClP_3 \rightleftharpoons RhClP_2 + P$ Ligand dissociation
2. $RhClP_3 \rightleftharpoons [RhClP_2]_2 + 2P$ Dissociative dimerization
3. $RhClP_2 + Un \rightleftharpoons RhCl(Un)P_2$ Olefin coordination
4. $[RhClP_2]_2 + 2Un \rightleftharpoons 2RhCl(Un)P_2$ Olefin coordination

5. $RhClP_3 + H_2 \rightleftharpoons RhH_2ClP_3$ Oxidative addition of H_2

6. $RhH_2ClP_3 \rightleftharpoons RhH_2ClP_2 + P$ Dissociation of P

7. $RhH_2ClP_3 + Un \rightleftharpoons RhH_2(Un)ClP_2 + P$ Coordination of Olefin

8. $RhH_2(Un)ClP_2 \rightleftharpoons RhH(R')ClP_2 \rightarrow R'H + RhClP_2$ Formation of hydrogenation product

where P = Ph_3P
Un = olefin
R' = alkyl

Dissociation equilibria of another important class of low-valent metal-phosphine complexes $M(PR_3)_n$ are as follows (M = Ni, Pd, and Pt):[33]

$$MP_4 \rightleftharpoons MP_3 + P \quad (1)$$

$$MP_3 \rightleftharpoons MP_2 + P \quad (2)$$

In the presence of unsaturated compounds equilibria 3 and 4 are also established.[34] Among these equilibria,

$$MP_3 + Un \rightleftharpoons MP_2(Un) + P \quad (3)$$

$$MP_2 + Un \rightleftharpoons MP_2(Un) \quad (4)$$

dissociative behavior of $Pt(PPh_3)_4$ has been examined to give the equilibrium constants that follow:

$$Pt(PPh_3)_4 \stackrel{K_1}{\rightleftharpoons} Pt(PPh_3)_3 + PPh_3 \quad K_1^{300°} = 1 M$$

$$Pt(PPh_3)_3 \stackrel{K_2}{\rightleftharpoons} Pt(PPh_3)_2 + PPh_3 \quad K_2^{300°} \simeq 10^{-6} M$$

The first equilibrium constant (K_1) shows extensive dissociation of $Pt(PPh_3)_4$ to $Pt(PPh_3)_3$ on dissolution in aromatic solvents. The dissociation is thought to be caused by the steric bulk of PPh_3. The second dissociation constant of $Pt(PPh_3)_4$ is quite small. However small the concentration of $Pt(PPh_3)_2$, its high reactivity overrides the concentration effect. This is manifested in its ready reaction with various π-acidic molecules such as CO, O_2, NO, and TCNE.[35]

$$Pt(NO)_2(PPh_3)_2 \xleftarrow{NO} [Pt(PPh_3)_2] \xrightarrow{2CO} Pt(CO)_2(PPh_3)_2$$

$$Pt(TCNE)(PPh_3)_2 \xleftarrow{TCNE} \quad \xrightarrow{O_2} Pt(O_2)(PPh_3)_2$$

[Pt(PPh$_3$)$_2$] has a very high catalytic activity in isomerizing *cis-trans*-azobenzene.[31] A very labile η^2-azobenzene complex has been proposed as an intermediate.

$$\text{Ph}\diagdown_{N=N}\diagup^{\text{Ph}} \xrightarrow{+\text{PtP}_2} \left[\begin{array}{c} \text{P}\diagdown_{\text{Pt}}\diagup^{\text{P}} \\ \diagup^{\text{N}-\text{N}}\diagdown \\ \text{Ph} \quad \text{Ph} \end{array}\right] \rightleftarrows$$

$$\left[\begin{array}{c} \text{P}\diagdown_{\text{Pt}}\diagup^{\text{P}} \\ \text{Ph}\diagup^{\text{N}-\text{N}}\diagdown \\ \text{Ph} \end{array}\right] \xrightarrow{-\text{PtP}_2} \text{Ph}\diagdown_{N=N}\diagdown^{\text{Ph}}$$

P = Ph$_3$P

When the bulkiness of the phosphine or phosphite ligands is increased, the extent of dissociation increases. The very bulky phosphine P(t-Bu)$_3$ yields π-acidic molecules (e.g., CO) but is inert to larger molecules (e.g., PhC≡CPh).[36] The bulkiness has been estimated semiquantitatively by the cone angles of the ligand.[37] Numerical values have been easily measured from molecular models of the free ligand as shown in Fig. 30, and typical values are as follows.

Phosphorus compound	Typical cone angles θ (degrees) (M = Ni)
P(CH$_3$)$_3$	118 ± 4
PPh$_3$	145 ± 2
P(OPh)$_3$	121 ± 10
P(cyclohex)$_3$	179 ± 10
P(t-Bu)$_3$	182 ± 10

It is important to note that the cone angles refer to the maximum steric requirement of the ligand and do not necessarily represent the real bulkiness of the coordinated state. For example, P(t-Bu)$_3$ has a cone angle of ∼180°, yet RhCl[P(t-Bu)$_3$]$_2$ can still coordinate one or two other small ligands.[38]

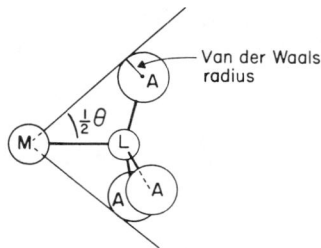

Fig. 30. The measurement of the ligand cone angle in the compound LA$_3$.

$$RhCl[P(t\text{-}Bu)_3]_2 \xrightarrow{H_2} RhH_2Cl[P(t\text{-}Bu)_3]_2$$
$$\downarrow CO$$
$$RhCl(CO)[P(t\text{-}Bu_3]_2$$

This property has also been well established with P(cyclohex)$_3$.

The variation of metals vertically in the periodic table has revealed an interesting trend in the coordinating tendency among isostructural complexes. In a series of phosphine complexes of the zerovalent nickel triad metals (Ni, Pd, Pt), the equilibrium constants for dissociation of C$_2$H$_4$ in M(C$_2$H$_4$)(PPh$_3$)$_2$ are in the order Pd > Pt > Ni. Since the metal-ethylene bond is largely governed by back-donation from the metal to C$_2$H$_4$, the observed trend follows a known trend in back-donation, Ni > Pt > Pd. Similar trends for back-donation in the vertical triad elements have been observed: Co > Ir > Rh and Fe > Os > Ru.[39]

Geometrical as well as electronic factors are important in influencing the η^2 coordination. When η^2 coordination to an electropositive metal center is dominated by σ-type donation from the olefin π_b orbital, *cis*-olefins coordinate more preferably than the *trans*-olefins do. However when the coordination is governed mainly by the back-donation to the π^* orbital, the preference is reversed, and the *trans*-olefins are favored. For example, *trans*-diimide HN=NH is calculated to be a stronger π acceptor from Ni(0) than is *cis*-diimide.[40]

The equilibrium constants shown in Table 8 vary over a wide range depending on the substituents attached to the C=C bond. Apparently electron-withdrawing substituents enhance the η^2-olefin coordination to zerovalent nickel. This trend indicates the importance of back-donation in the nickel—η^2-olefin bond. Apart from the electronic effect, bulkiness of the substituents also influences the K_1 values. Figure 31 gives a scheme for interaction of vacant π^* orbitals of olefins with filled metal d orbitals.[41]

Table 8. Selected Examples of Equilibrium Constants in NiL$_3$–Olefin Systems

$$\text{NiL}_3 + \text{olefin} \xrightleftharpoons{K_1} \text{NiL}_2(\text{olefin}) + \text{L}$$

(L = P(*O-o*-tolyl)$_3$)

Olefin	K_1
trans-NC-CH=CH-CN	1.6×10^8
CH$_2$=CH-CN	4.0×10^4
CH$_2$=CH-Ph	1.0×10^1
cis-Et-CH=CH-Et	2.7×10^{-3}
Et$_2$C=CH$_2$	2.3×10^{-3}

In summary, the coordination equilibria of labile metal-olefin or related complexes are determined by a delicate balance of steric and electronic effects. Further extensive investigation is needed to clarify the complex equilibria in real catalyst systems.

2.2. Oxidative Addition and Reductive Elimination

It has long been known that square planar Pt(II) complexes PtX$_2$L$_2$ add a halogen molecule to give corresponding octahedral Pt(IV) complexes PtX$_4$L$_2$. With the development of the chemistry of transition metal carbonyls and phosphine complexes in recent years, a number of similar addition reactions to low-valent metal complexes, such as Fe(O), Ru(O), Co(I), Rh(I), and Ir(I),

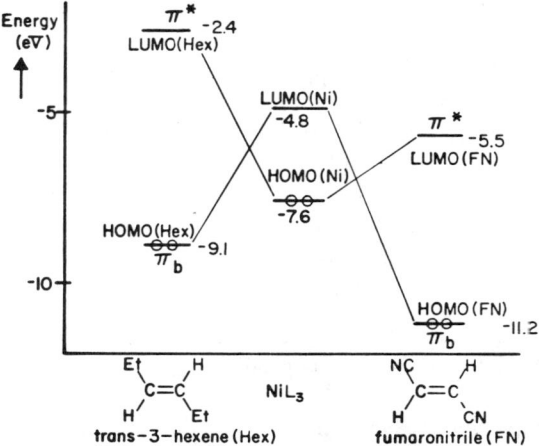

Fig. 31. Orbital interactions in olefin coordination to Ni(0).

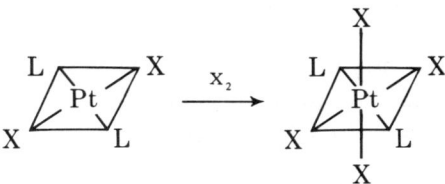

have been reported and have been termed "oxidative addition."[39,42] A characteristic feature of oxidative addition is that the formal oxidation state of the metal is raised by 2 units. For example, Vaska's complex $IrX(CO)(PPh_3)_2$ shows a remarkable ability to add small molecules such as H_2, HX, O_2, or X_2 to give octahedral Ir(III) complexes. This discovery in 1962 prompted further research

on similar reactions between many other metal complexes and small molecules. As a result, numerous metal complexes [e.g., $Pt(PPh_3)_4$, $RhCl(PPh_3)_3$, and $Fe(CO)_3(PPh_3)_2$] have been found to undergo oxidative addition. Table 9 gives some typical examples, as well as the structure of the resulting oxidative adducts.

Table 9. Metal Complexes and Their Oxidative Adducts

Metal Complexes		Oxidative Adducts
General Formula[a]	Example	with H_2 or Br_2
trans-$M^I XLL'_2$	$IrCl(CO)(PPh_3)_2$	$IrH_2Cl(CO)(PPh_3)_2$
$M^I XL_3$	$RhCl(PPh_3)_3$	$RhH_2Cl(PPh_3)_3$
$M^{II} X_2 L_2$	$PtI_2(PBu_3)_2$	$PtBr_2I_2(PBu_3)_2$
$M^0 L_2$	$Pt[P(cyclohex)_3]_2$	trans-$PtH_2[P(cyclohex)_3]_2$
$M^0 L_3$	$Pd(PPh_3)_3$	trans-$PdBr_2(PPh_3)_2$
ML_4	$Ni(CN$-t-$Bu)_4$	$NiBr_2(CN$-t-$Bu)_2$
$ML_2 L'_3$	$Ru(CO)_3(PPh_3)_2$	$RuBr_2(CO)_2(PPh_3)_2$

[a] X = anionic ligand; L or L' = neutral ligand.

The adding molecules (usually called addenda) have been roughly classified into two types: one has been characterized as being split into two η^1 ligands (formally anionic); the other functions as a η^2 ligand without being split. Table 10 gives typical examples of both types.

The wide applicability and chemical significance of oxidative addition in the field of inorganic chemistry became recognized in a surprisingly short period of time, and now it is one of the most important elementary reactions in inorganic chemistry. Many catalytically active transition metal complexes participate in

Table 10. Examples of Oxidative Addenda

$2 \times \eta^1$ Addenda: Single-Bond Breaking on Reaction	
H—H	H—X (X = Cl, Br, I, CN, SCN)
H—ER	(E = O, N, S; R = H, CH_3, etc.)
H—C≡CR	H—C≡C, H—C<
H—MR_3	(M = Si, Ge, Sn; R = H, CH_3, etc.)
X—R	(X = Cl, Br, I; R = organic substrate)
X—ML_n	(M = Hg, Au, etc.); L = ligand (neutral or anionic)
η^2 Addenda: Forming η^2-Coordination to Metal	
O=O, S=S	
O=N—R, S=N—R	
R—N=N—R, R—N=C=O, R—N=C=S	
O=C=O, S=C=O, S=C=S	

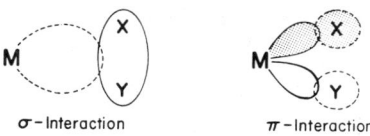

Fig. 32. Interactions between metal orbitals and relevant orbitals in X—Y molecules.

oxidative addition. For example, $RhCl(PPh_3)_3$ adds small molecules such as H_2, O_2, or CH_3I, and these small molecules are then "activated" for further reactions. Sometimes oxidative addition is a key step in homogeneous catalysis.[42]

There are several important factors governing oxidative addition. The electronic properties of the addenda, as well as the energy levels and orbital symmetry properties of their LUMOs, are in most cases crucial factors. This is because the addenda (X—Y) with the LUMOs in relevant levels can interact with filled metal orbitals (HOMOs), resulting in the addition with electron flow from the metal to the addenda. When the symmetry of the LUMO (σ or π) is such that it has a positive overlap with filled metal d orbitals in a π manner as shown in Fig. 32, the relative levels of these orbitals determine the feasibility of oxidative addition. For example, among many square tetracoordinated d^8 metal complexes, the trend favoring oxidative addition is $Ir(I) > Pt(II) \gg Au(III)$; that is, metals with filled d orbitals at higher levels tend to oxidatively add small molecules (see Table 11). The same trend have also been observed with five-coordinate d^8 metal species, $Fe(O) > Co(I) > Ni(II)$.[39]

The energy level of the LUMO of a small molecule (X—Y) varies with the electronic properties of the participating atoms (X and Y) and with the substituents attached to them. The lower the LUMO level, the stronger the interaction with a low-valent metal when positive overlap occurs with relevant metal orbitals. Thus halogens (X_2: X = Cl, Br, I) generally oxidatively add to the metal complexes more easily than H_2. The oxidative addition of saturated C—H bonds is more difficult than that of H_2. When small molecules containing an acidic hydrogen (H—X) have been allowed to react, ionic dissociation of the

Table 11. The Trend in Oxidative Addition of Square d^8 Complexes

Reactant	Products for Reaction with O_2 and Cl_2	
	O_2	Cl_2
$Ir^I Cl(CO)(PPh_3)_2$	$IrCl(CO)(O_2)(PPh_3)_2$	$IrCl_3(CO)(PPh_3)_2$
$Pt^{II}Cl_2(PPh_3)_2$	—	$PtCl_4(PPh_3)_2$
$Au^{III}Cl_3(PPh_3)_2$	—	—

Table 12. A Comparison of Kinetic Rate Constants and Parameters for Oxidative Addition of O_2 and H_2 to $[M(cis\text{-}Ph_2P\text{---}CH\text{=}CH\text{---}PPh_2)_2]^+$ BPh_4^- in PhCl at 25°

Addendum	M	Second-order rate constants of oxidative addition step (M^{-1} sec^{-1})	ΔH^* (kcal/mol)	ΔS^* (eu)	ΔG^* (kcal/mol)
O_2	Co	1.7×10^4	3.4	−28	10.3
O_2	Rh	0.12	11.6	−24	18.8
O_2	Ir	0.47	6.5	−38	17.8
H_2	Co	1.2×10^5	3.6	−23	11.0
H_2	Ir	6.7×10^3	5.0	−24	12.2

L. Vaska, L. S. Chen, and W. V. Miller, *J. Am. Chem. Soc.*, 93, 6671 (1971).

molecules giving a proton, and the conjugate anion is an important step preceding the addition. The protonation of the metal ion in a low-valent complex yields a hydridometal cation, which then coordinates with the conjugate anion with or without dissociation of a neutral ligand (L) to complete the oxidative addition (also see p. 123).

$$HX \rightleftharpoons H^+ + X^- \qquad X^-: \text{Conjugate anion}$$
$$H^+ + ML \rightleftharpoons [HML_n]^+ \qquad \text{Protonation step}$$
$$[HML_n]^+ + X^- \rightleftharpoons HMXL_n \text{ or } n-1 \qquad \text{Coordination of anion}$$

Since oxidative addition is a thermally allowed reversible process, an equilibrium is attained in many cases. The rate and position of the equilibrium are delicately influenced by intricate factors that are largely unknown at present. Table 12 lists some examples of the kinetic rate constant and parameters.

In some cases dissociation of a neutral ligand molecule critically determines the ease of addition because the dissociation (preequilibrium) provides a reactive

$$ML_n \rightleftharpoons ML_{n-1} + L \qquad \text{Preequilibrium for oxidative addition}$$

coordinatively unsaturated metal species. Oxidative addition to coordinatively saturated molecules such as $Fe(CO)_5$, $HCo(N_2)(PPh_3)_3$, or $Ni(PPh_3)_4$ involves such predissociation equilibria. For example,

$$HCo(N_2)(PPh_3)_3 \rightleftharpoons N_2 + HCo(PPh_3)_3$$

The energetics of oxidative addition depend not only on the energies E of the participating bond-breaking or bond-making process but also on the promotion energy P, which is a measure of energy change associated with an increase in coordination number and the oxidation state of the metal. To shift the equilibrium of any oxidative addition to the right, the following relationship, where E_{MZ}, E_{MY}, and E_{YZ} are the bond energies for M—Z, M—Y, and Y—Z, respectively, must be fulfilled:

Energy Relationship for Oxidative Addition

$$(M + YZ \rightarrow Y-M-Z)$$

$$E_{MZ} + E_{MY} \geq E_{YZ} + P$$

Among these energy values, only E_{YZ} is known in most cases. Therefore this relationship is of limited value in predicting a reaction.

The stereochemistry of oxidative addition is also important. The addition of a hydrogen molecule to Vaska's complex {IrCl(CO)(PPh$_3$)$_2$} occurs in a cis manner, implying an essentially concerted bimolecular mechanism. Consideration of the symmetry of relevant molecular orbitals of participating molecules suggests an orbital interaction scheme (Fig. 33), where the filled $d\pi$ orbital of Ir^I and the vacant σ^*_{H-H} orbital of a hydrogen molecule overlap in a positive manner to stabilize the bimolecular transition state of the reaction. Another important orbital interaction occurs between the vacant p orbital of Ir^I and σ_b orbital of H$_2$ (Fig. 34). Since a filled $d\pi$ orbital is lower in energy than that of σ^*_{H-H} orbital, a higher $d\pi$ level would accelerate the oxidative addition of H$_2$. In fact, IrF(CO)(PPh$_3$)$_2$ adds an H$_2$ molecule more readily than IrCl(CO)(PPh$_3$)$_2$ or IrBr(CO)(PPh$_3$)$_2$. The fluoride ligand probably acts as an effective π-donating auxiliary ligand, and the π basicity at the metal is thereby increased, while the σ basicity is decreased.[43]

Oxidative addition, which was discussed in connection with Tables 11 and

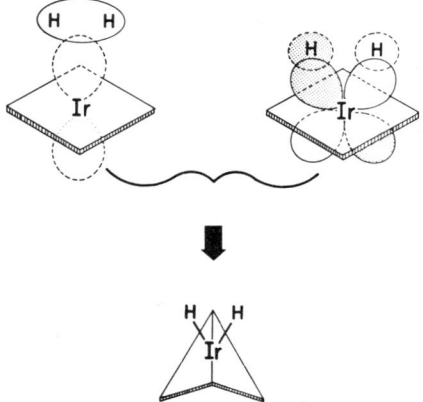

Fig. 33. The oxidative addition of H$_2$ to a square planar Ir^I complex.

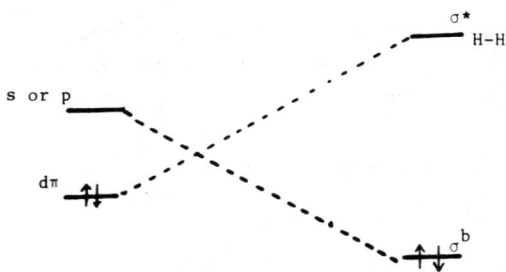

Fig. 34. An orbital diagram showing interaction between the frontier orbitals of Ir^I and H_2.

12, of typical protic acids (e.g., HBr) proceeds exclusively in a cis manner when it is allowed to react with $IrCl(CO)(PPh_3)_2$ in *nonpolar* solvents. Although this result implies a concerted mechanism, the H—Br bond cannot add to the metal synchronously because of its highly polar character. As a result, HBr addition in *polar* organic solvents (e.g., DMF) occurs in a trans manner. A stepwise mechanism has been suggested: incipient protonation to the metal followed by addition of Br^- by a trans attack.[44]

$$\underset{\substack{\text{H}\\ \text{P}\diagup\text{Ir}\diagdown\text{Cl}\\ \text{OC}\diagup\diagdown\text{P}\\ \text{Br}}}{} \xleftarrow{\text{HBr in DMF}} \underset{\substack{\text{P}\diagup\text{Ir}\diagdown\text{Cl}\\ \text{OC}\text{P}}}{} \xrightarrow{\text{HBr in nonpolar solvent}} \underset{\substack{\text{Br}\\ \text{P}\diagup\text{Ir}\diagdown\text{H}\\ \text{OC}\diagup\diagdown\text{P}\\ \text{Cl}}}{}$$

$P = PPh_3$

Alkyl halides are one of the most important addenda in oxidative addition reactions because they give rise to reactive M—C bonds that become active sites in many catalytic reactions. The mechanism of the addition has been a subject of a considerable dispute in recent years. Kinetic studies concerning the reaction between CH_3X and $IrX(CO)(PPh_3)_2$ have revealed an S_N2 mechanism, and Ir^I is suspected to behave as a nucleophile in a manner similar to organic amines.

Inversion at the carbon has been expected.[45]

The stereochemical fate of the chiral α carbon in optically active alkyl halides

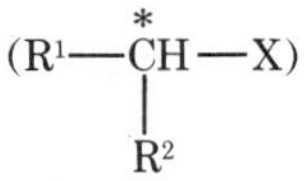

$$(R^1-\overset{*}{\underset{R^2}{C}H}-X)$$

after oxidative addition could give a decisive answer to this problem. Therefore optically active halides, such as the one shown below for $Pd(PPh_3)_4$, have been examined for oxidative addition to $M(L)_n$.[46] In the case illustrated the trapping experiment with CO in insertion gave an optically active adduct, which has an inverted configuration at the chiral α carbon. When the trapping is not done, the optical activity of the oxidative addition product is lost. This phenomenon may be due to the reversibility of the oxidative addition.

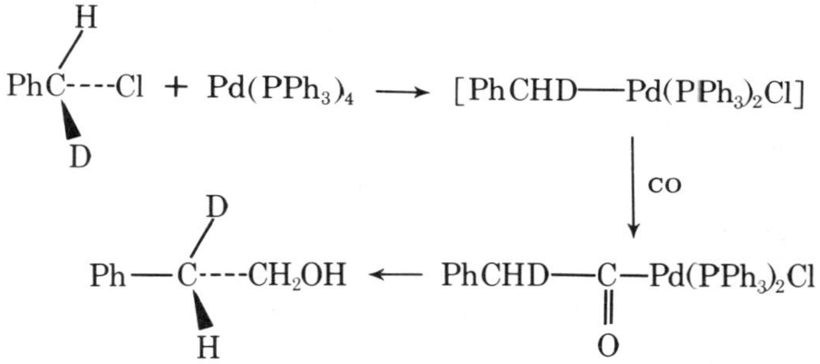

$$PhC(H)(D)\text{----}Cl + Pd(PPh_3)_4 \longrightarrow [PhCHD\text{----}Pd(PPh_3)_2Cl]$$

$$\downarrow CO$$

$$Ph-C(D)(H)\text{----}CH_2OH \longleftarrow PhCHD-\underset{O}{\overset{\parallel}{C}}-Pd(PPh_3)_2Cl$$

The oxidative addition of various organic halides to $IrCl(CO)(PR_3)_2$ or $Pt(PPh_3)_3$ has sometimes been effectively hindered by the addition of a small amount of free-radical trapping reagents.[47] Thus addition of duroquinone to the reaction mixture $RX/Pt(PPh_3)_3$ inhibited the reaction completely. This observation indicates that the reaction has a radical-chain character. Supporting this concept is the remarkable accelerating effect of light on the reaction of $CH_3CHCl-CO_2Et$ with $[Rh(CNBu)_4]^+$. Here $[Rh^{II}X(CNBu)_4]^+$ is the probable chain carrier.[48]

Equations 5 through 8 depict a proposed radical-chain mechanism.

$$RX + ML_n \rightarrow R\cdot + X\dot{M}L_n \tag{5}$$

$$X\dot{M}L_n + RX \rightarrow XM(R)L_m + X\cdot \tag{6}$$

$$\text{or } X\dot{M}L_n + RX\text{----}ML_n \rightarrow XM(R)L_m + X\dot{M}L_n \tag{7}$$

$$X\cdot + ML_n \rightarrow X\dot{M}L_n \tag{8}$$

Since there are a wide variety of combinations of low-valent metal complexes with different alkyl or alkenyl halides, it is difficult to draw any general mechanistic conclusions at this time.

Reductive elimination has been regarded as the reverse of oxidative addition, and it is also one of the important elementary reactions in catalysis. The following examples illustrate reductive elimination steps in a hydrogenation and a C—C coupling reaction catalyzed by transition metal complexes.

1. Last step of hydrogenation:

$$M\begin{array}{c}CH_2R\\H\end{array} \longrightarrow M + \begin{array}{c}CH_2R\\|\\H\end{array}$$

2. C—C Bond-forming step in a C—C coupling reaction:

$$M\begin{array}{c}CH_2R\\CH_2R\end{array} \longrightarrow M + \begin{array}{c}CH_2-R\\|\\CH_2-R\end{array}$$

When metal-hydrogen or metal-carbon bonds utilize only s or p metal orbitals, these eliminations, if synchronous, would be thermally forbidden reactions as predicted by the orbital symmetry conservation rules. Actually, however, a synchronous elimination mechanism (concerted mechanism) has been proposed on the basis of the observed retention in stereochemistry of the α carbon of the leaving group R.[49]

$$Cp_2Mo\begin{array}{c}H\\C \blacktriangleleft CO_2CH_3\\CHDCO_2CH_3\\D\end{array} \xrightarrow[\text{in toluene}]{\text{room temp.}} [Cp_2Mo] + \begin{array}{c}H\\C\\D\end{array}\begin{array}{c}CO_2CH_3\\CHDCO_2CH_3\end{array}$$

Elimination of n-butane or biphenyl occurs readily with thermal decomposition of diethyl(dipyridyl)nickel or diphenylbis(triphenylphosphine)nickel, respectively.[50] These geminal eliminations have been explained by a concerted mechanism. The participation of d orbitals in the metal-hydrogen or metal-carbon bond is one of the characteristics of organo transition metal complexes.

Energetic conditions for a reductive elimination are just the opposite of those for the corresponding oxidative addition. Thus an energetically stable X—Y bond formation together with a high promotion energy (P) of a ML_n species will contribute to the reductive elimination of X—Y from $L_nM\begin{smallmatrix}X\\Y\end{smallmatrix}$. When the reduced state L_nM is stabilized with some π-acidic ligands, the promotion

$$L_nM\overset{X}{\underset{Y}{\diagdown}} \xrightarrow{\text{reductive elimination}} L_nM + X\text{—}Y$$

energy $(P)^*$ will be higher, and reductive elimination is facilitated. For example, substitution of tricyclohexylphosphine ligands in trans-$R_2Ni[P(cyclohex)_3]_2$[51] by a π-acidic triarylphosphite results in the reductive coupling of R groups together with the formation of $Ni[P(OAr)_3]_4$. Similar reductive alkyl couplings have also been observed in the chemistry of alkyl compounds of some Group V metals.

$$SbPh_5 \rightarrow SbPh_3 + Ph\text{—}Ph$$

This reaction can be explained by the high promotion energy of Sb^{III} to Sb^V and also by a mixing of d character into the Sb—C bonds.

A detailed kinetic study by Yamamoto et al.[52] on the thermal elimination of alkanes from cis-dialkyl(dipyridyl)nickel has revealed that coordination of π-acidic ligands, such as acrylonitrile, remarkably accelerates the reaction. This effect has been ascribed to the "activation" of Ni—C bonds by π coordination. A similar "activation" is probably responsible for the rapid olefin insertion into an M—C bond when the olefin is π-coordinated in an adjacent (cis) position.

The nature of the auxiliary ligands is also one of the important factors affecting oxidative addition or reductive elimination. In general, chelating ligands such as dipyridyl or chelating diphosphines tend to stabilize the oxidized state as can be seen from the following examples. The high trans effect of the alkyl group labilizes the trans isomer.

$$\text{Py-Ni}(C_2H_5)_2(\text{Py})$$
Unstable at room temperature

$$\text{(bipy)Ni}(C_2H_5)_2$$
Stable up to 80°

$$\text{(MeP}_3)(PMe_3)\text{Ni}(CH_3)_2$$
Decomposes 69 to 71°

$$\text{Me}_2\text{PCH}_2CH_2\text{PMe}_2\text{-Ni}(CH_3)_2$$
Decomposes at 130°

Electron-donating bulky diphosphines have recently been found to be especially effective in stabilizing otherwise very labile dihydridoplatinum species.[53] Chelation is another important factor for stabilization of the labile PtH_2 moiety as exemplified by $PtH_2[(t\text{-Bu})_2PCH_2CH_2P(t\text{-Bu})_2]$. Chelation also forces a cis geometry for the addenda.

$$\text{trans-}(Ph_3P)_2Pt(H)_2$$
Too unstable to be isolated

$$\text{trans-}((Cyclohex)_3P)_2Pt(H)_2$$
Stable up to 158–159° in air with decomposition

$$(t\text{-Bu})_2PCH_2CH_2P(t\text{-Bu})_2\text{-Pt}(H)_2$$
Stable in air, m.p. 224°–226° with decomposition

When triad elements (e.g., Ni, Pd, Pt) of transition metals are compared in terms of reductive elimination, for example, among some isostructural aryl complexes, the ease of this process decreases in a series, Ni, Pd, Pt. This trend has been related to the thermal stability of the metal-carbon σ bond in these complexes.[54]

2.3. Insertion and De-insertion

Insertion of unsaturated compounds into M—H or M—C bonds is an important elementary reaction in various types of homogeneous catalysis. In general, this reaction can be represented as shown below, where M—Z is M—H, M—C, or M—N; X=Y is C=C, C=N, N=N, and so on, and :X—Y is :C=O, :C≡N—R, or :CR$_2$, and so forth

$$M-Z \xrightarrow{X=Y} M-X-Y-Z$$
$$M-Z \xrightarrow{:X-Y} M-X(Y)-Z$$

On consideration of the inserting molecules, the same reaction has been regarded as a 1,2-addition of the M—Z bond to the X=Y linkage or a 1,1-addition of the M—Z bond to the X atom of the :X—Y molecule. The insertion into the M—H or the M—C bond is called "hydrometallation" or "carbometallation," respectively.

Since there is a great variety of combinations between the M—Z bond and X=Y or :X—Y molecules, a comprehensive coverage of the whole scope of this reaction is not attempted here. Some typical examples are explained below.

1. *Insertion into polar M—Z bonds.* The M—H bonds in hydrides of aluminum or boron have polar characters that induce polarized unsaturated bonds to insert into the M—H bond. The insertion of a ketonic C=O bond into an Al—H bond is illustrated below.

$$[R_2AlH] + R_2C=O \longrightarrow R_2Al-O-CHR_2$$

Even nonpolar olefinic double bonds insert into Al—H bonds to give alkylaluminum compounds. The following intermediate with a polarized C=C bond has been proposed:

$$[R_2AlH] + H_2C{=}CH_2 \longrightarrow \begin{bmatrix} \overset{\delta-}{H_2C}{\text{---}}\overset{\delta+}{CH_2} \\ \vdots \quad \vdots \\ R_2Al{\text{---}}H^{\delta-} \end{bmatrix}$$

$$\downarrow$$

$$R_2Al{\text{---}}CH_2{\text{---}}CH_3$$

$$R_2Al(CH_2{\text{---}}CH_2)_n CH_2CH_3 \xleftarrow{\text{further insertion of } C_2H_4}$$

This reaction forms the basis for the manufacture (Ziegler process) of straight chain higher alcohols from C_2H_4 and "$AlEt_3$." Similar insertions of alkenes into the B—H bonds of diborane (B_2H_6) proceed remarkably fast at ambient temperature in polar organic solvents such as diglyme.[55] The strongly accepting character of the boron center seems to induce the polar insertion into the polarized B—H bond. In the gas phase the same insertion is a very slow process even at elevated temperature (e.g., at 130°). Rapid and convenient preparative methods, known widely as "hydroboration," have been based on this reaction.

"B_2H_6" + $\diagup$C=C$\diagdown$ $\xrightarrow{\text{in THF}}$ B(—C—CH)$_3$ $\xrightarrow[OH^-]{H_2O_2}$ HO—C—CH

$$\left(\text{e.g. } NaBH_4 + BF_3{\cdot}Et_2O + \text{[alkene]} \xrightarrow[OH^-]{H_2O_2} \text{[alcohol]} \right)$$

2. *Insertion into weakly polar M—Z bond.* The M—H or M—C bonds in transition metal complexes are often essentially covalent or only weakly polar. Despite this covalent character, insertion of olefins into some transition metal–hydrogen or –carbon bonds proceeds quite rapidly. In particular, labile hydrido complexes of d^1, d^2 or d^8, d^9 complexes [e.g., ZrH_2Cp_2, $ZrHClCp_2$, or $NiH(X)(PR_3)_2$] readily insert olefins to give labile metal alkyl complexes.[56] In most cases the insertion into an M—H bond can be reversed when a metal alkyl is heated, yielding olefins and metal hydrides because the insertion and

$$M{\text{---}}H + \diagup C{=}C \diagdown \rightleftharpoons \diagup\overset{}{C}{\text{---}}\overset{}{C}\diagdown_{\underset{M \quad H}{}}$$

de-insertion (β-hydride elimination) are both thermally allowed, reversible processes.

The insertion of olefins into transition metal–hydrogen bonds forms the basis for many important homogeneous olefin catalyses such as hydrogenation, isomerization, polymerization, and carbonylation. The stereochemistry of insertion of olefins into M—H or M—C bonds is of fundamental importance in deciding stereochemical features of hydrogenation or polymerization. For example, exclusive cis addition of hydrogen to the olefinic double bond occurs by catalysis utilizing Wilkinson's rhodium complex $RhCl(PPh_3)_3$.[57] Therefore olefin cis insertion into one of the Rh—H bonds has been proposed. A similar cis insertion also occurs in the case of the hydrogenation of an acetylenic bond. These

stereochemical results are also valid for many other transition metal complex catalysts. The intermediacy of a hydrido- σ-alkyl metal complex has been verified in the case of olefin hydrogenation with MoH_2Cp_2 catalyst.[49] Here careful examination of an equimolar reaction mixture of dimethyl fumarate and MoD_2Cp_2 by ^1H nmr spectroscopy has revealed formation of a thermally labile, air-sensitive deuterioalkyl complex that slowly liberates racemic dimethyl 1,2-dideuterosuccinate on standing at 25°. The following mechanism has been proposed (Fig. 35). The intermediate is a thermally excited parallel metallocene structure, which has the important η^2 coordination of the electronegative olefin leading to cis insertion.

The cis insertion is formally a $(2_s + 2_s)$ reaction if the M—H bond can be regarded as a covalent two-electron system. Woodward-Hoffmann rules have predicted this cycloaddition reaction to be a thermally forbidden process. Actually however, the insertion proceeds very rapidly via a thermal process, in some

Four-centered transition state

cases with typical covalent transition metal-hydrogen bonds. Participation of

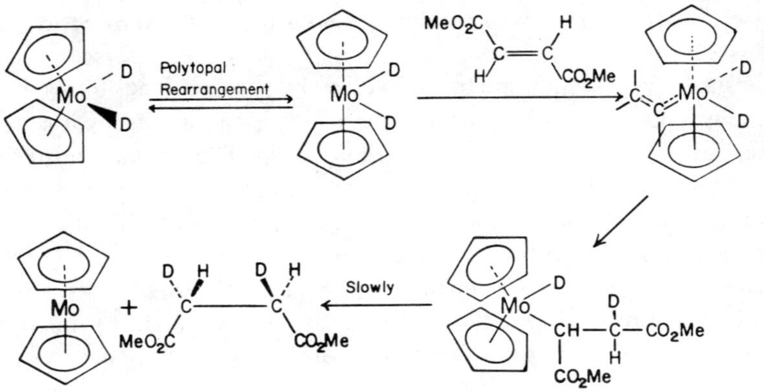

Fig. 35. A proposed scheme for olefin insertion into MoD_2Cp_2.

d orbitals in the M—H bonding seems to lower the energy barrier created by the symmetry-forbidden process. The role of the *d* orbitals in matching the orbital symmetry for concerted reaction has been stressed.[49] Since thermal insertion is reversible, the same orbital interaction is also involved in the β elimination (de-insertion) of transition metal alkyls having β hydrogens. The stereochemistry of insertion into M—C bonds has also been studied using deu-

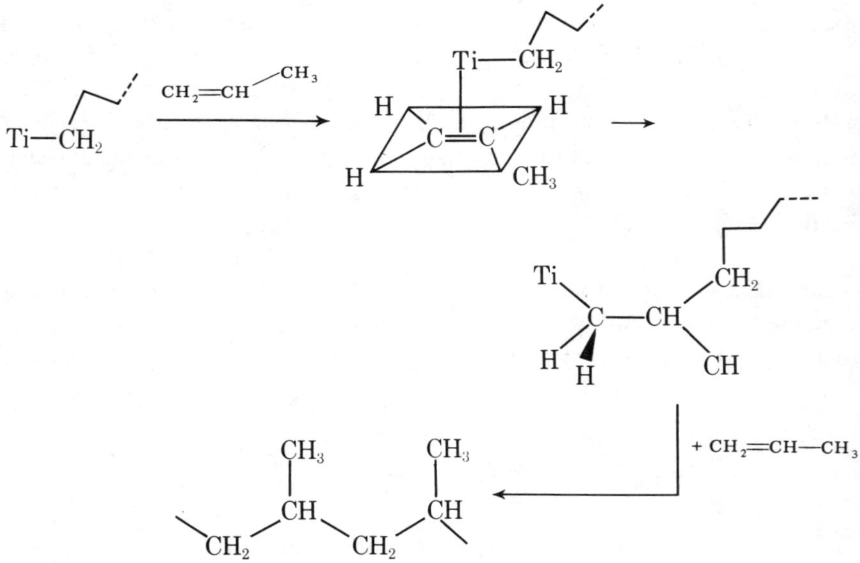

Stereoregular polypropylene

Elementary Reactions in Transition Metal Chemistry

terated ethylenes or propylenes.[58] The reaction is stereospecific for cis insertion. On the basis of the observed stereochemistry of "stereospecific high polymerization" of propylene by Ziegler-Natta catalysts $Et_3Al/TiCl_3$, a mechanism involving essentially concerted cis insertion with highly effective steric control has been proposed. The orbital interaction is probably similar to that shown above for the insertion into an M—H bond.

Some examples of the insertion of carbenic species such as CO, CNR, or $:CH_2$ into M—H or M—C bonds are illustrated below:

$$CH_3Mn(CO)_5 + CO \longrightarrow CH_3COMn(CO)_4 \qquad \text{ref. 59}$$

$$CH_3NiI(CNR)_2 + 2CNR \longrightarrow \underset{\underset{R-N}{RNC}}{\overset{\underset{NR}{I}}{Ni}}\overset{C-CH_3}{\underset{N-R}{C-C}} \qquad \text{ref. 60}$$

$$HMn(CO)_5 + CH_2: \longrightarrow CH_3Mn(CO)_5 \qquad \text{ref. 61}$$
$$(\text{from } CH_2N_2)$$

These reactions, which have been called carbenic insertions, are important elementary reactions in catalytic carbonylations, polymerizations of CNR, and so on. A three-centered transition state has been proposed for these reactions based on the stereochemistry and kinetic characteristics. Since the CO reaction is reversible, decarbonylation of aldehydes probably proceeds through a

$$R-\underset{O}{\overset{H}{\underset{\|}{C}}} \rightleftharpoons R-\underset{O}{\overset{\|}{\underset{\|}{C}}}-M-H \rightleftharpoons R-M-H + CO$$

$$L_nM\overset{R}{\underset{C\equiv O}{}} \rightleftharpoons \left[L_nM\overset{R}{\underset{C\equiv O}{\cdots}}\right] \rightleftharpoons L_nM-\underset{O}{\overset{\|}{C}}-R$$

similar three-centered transition state. The reversible nature of carbonylation (CO insertion) has been best illustrated by a recent example: an equilibrium mixture resulted under 1 atm of CO at 20°C.[62]

If successive CO insertions occur, a polyketone

$$(-\underset{\underset{O}{\|}}{C}-\underset{\underset{O}{\|}}{C}-\underset{\underset{O}{\|}}{C}-)_x$$

will result. However, this has not been realized yet. Isocyanides are more carbenic in some of their reactions, and successive insertions take place with relevant nickel catalysts (cf. $[\eta\text{-}^3C_3H_5NiCl]_2$).[63]

Carbenes are much more reactive and insert into various sites on the other molecule. The most reactive carbene, CH_2, readily inserts into M—H, M—C, as well as M—Cl bonds.[64] Some nickel complexes catalyze formation of polymethylene.

$$CH_2N_2 \xrightarrow{IrCl(CO)(PPh_3)_2} Ir(CH_2Cl)(CO)(PPh_3)_2$$

$$CH_2N_2 \xrightarrow{NiCp_2 \text{ or } BF_3} (-CH_2-)_n$$

The stereochemistry of CO insertion into an M—C bond has been studied with an isotopically labeled metal carbonyl $CH_3{}^{13}COMn(CO)_5$.[65a] Infrared spectroscopic analysis of the CO elimination products after the thermal reaction has revealed the presence of $CH_3Mn(CO)_4(cis\text{-}^{13}CO)$. This result means that the CH_3 has migrated. The CO group cis to the CH_3 has inserted into the Mn—CH_3 bond. Therefore the CO insertion has been usually regarded as "methyl migration" to one of the neighboring CO groups. It is important to note that coordinating nucleophiles (e.g., PPh_3) also induce CO insertion. For CO insertion to an Ir—C bond, a somewhat different mechanism has been proposed.[65b]

$$CH_3Mn(CO)_5 + PPh_3 \longrightarrow CH_3\underset{\underset{O}{\|}}{C}-Mn(CO)_4(PPh_3)$$

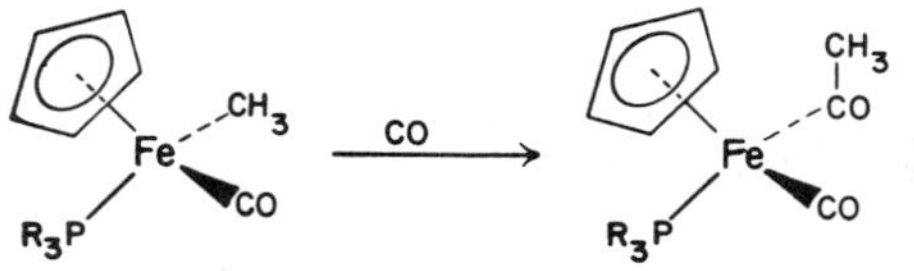

Scheme 5

The stereochemistry at the metal was investigated utilizing optically active pseudotetrahedral iron complexes. As Scheme 5 shows, the reaction proceeds with *retention* of configuration at the chiral iron atom.[66] When a chiral σ carbon

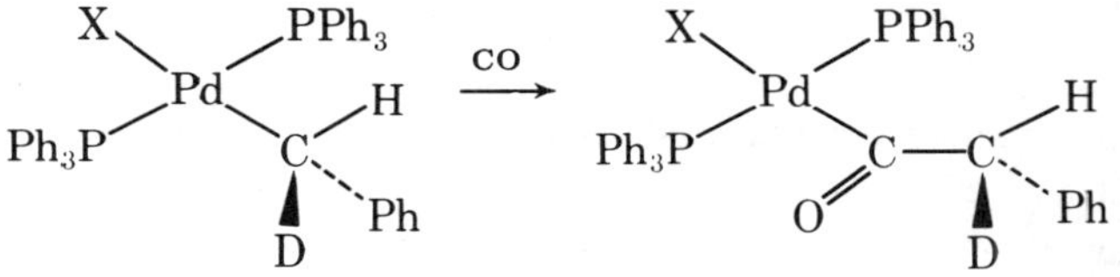

has directly attached to a metal, CO insertion also yielded the corresponding acyl complex with *retention* at the α-carbon atom.[67] These results give strong support to a mechanism involving an essentially nonpolar, three-center transition state.

2.4. Cycloadditions and Electrocyclic Reactions

In cycloadditions such as the Diels-Alder reaction (Fig. 36) two or more covalent (σ or π) bonds interact to give a cyclic product through a cyclic nonpolar transition state.[68] Since four π electrons of butadiene and two π electrons of acrylate are involved, this cycloaddition is designated as a $(4_s + 2_s)$ reaction. The participation of a metal in this cycloaddition at any state exercises a great influence on rate, stereochemistry, and selectivity.

An electrocyclic reaction has been defined as the formation of a single bond

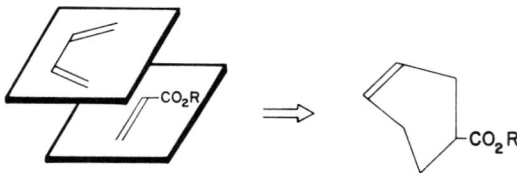

Fig. 36. A typical example of cycloaddition (the Diels-Alder reaction).

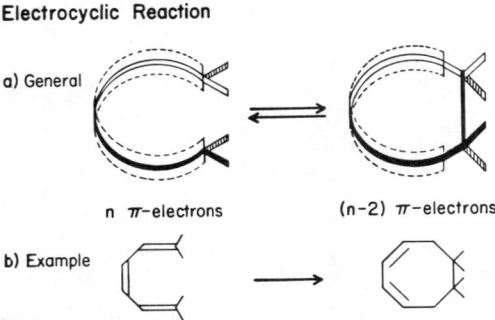

Fig. 37. An electrocyclic reaction involving n π electrons.

between the termini of a conjugated polyene, and cycloreversion has been defined as the reverse reaction. These reactions have been well established and have been useful reactions because of their unique stereoselectivity in the construction of certain organic structures. An example appears in Fig. 37. Cycloaddition and electrocyclic reactions both belong to a more general category of organic reactions called "pericyclic reactions," which occur through neither polar nor radical intermediates but through electronic reorganization around the participating σ or π bonds in a cyclic array. The well-known Woodward-Hoffmann rules can predict the feasibility as well as the stereochemistry of these "pericyclic reactions." Since metal-carbon or metal-hydrogen bonds in many organo transition metal complexes are essentially covalent, the application of the Woodward-Hoffmann rules to reactions involving these bonds and other covalent bonds has recently become very important.

Some examples of cycloaddition in the organometallic field are illustrated below.

1. Cluster formation by the cyclodimerization of two M≡M bonds:

$$2(OC)_3Co\equiv Co(CO)_3 \longrightarrow (OC)_3Co\underset{Co(CO)_3}{\overset{Co(CO)_3}{\diamond}}Co(CO)_3$$

2. Bridged acetylene complex formation by cycloaddition between M≡M and C≡C in a $(\pi 2_a + \pi 2_s)$ fashion (where, a and s designates an antarafacial and suprafacial reactions, respectively):

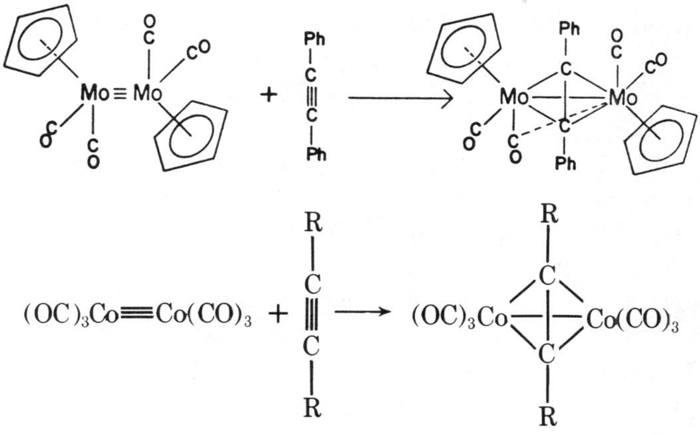

Scheme 6

Many other analogous reactions of this type are known, as Scheme 6 suggests.[69,70]

Carbyne complexes $L_nM{\equiv}C{-}R$ have been also expected to exhibit similar reactions (e.g., cyclodimerization or cycloaddition with acetylene).

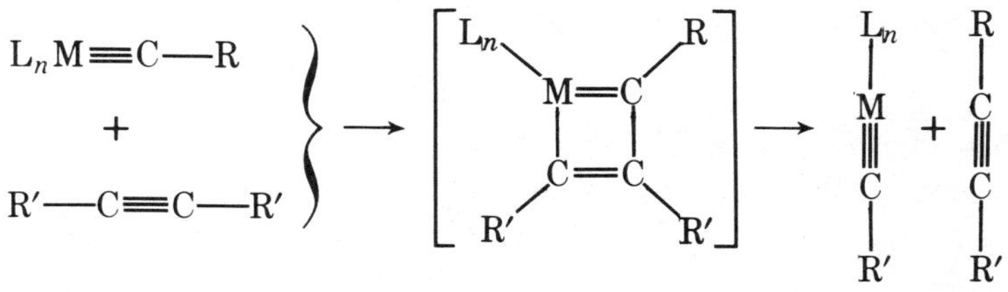

This reaction illustrates a possible pathway of catalyzed alkyne disproportionation by some metathesis catalysts ($R{-}C{\equiv}C{-}R + R'C{\equiv}CR' \leftrightarrows R{-}C{\equiv}C{-}R'$).

The reaction of carbyne complex with the $M{\equiv}M$ bond of some relevant binuclear complexes yields a trimetallobicyclobutane structure that has also been prepared by different routes.[71]

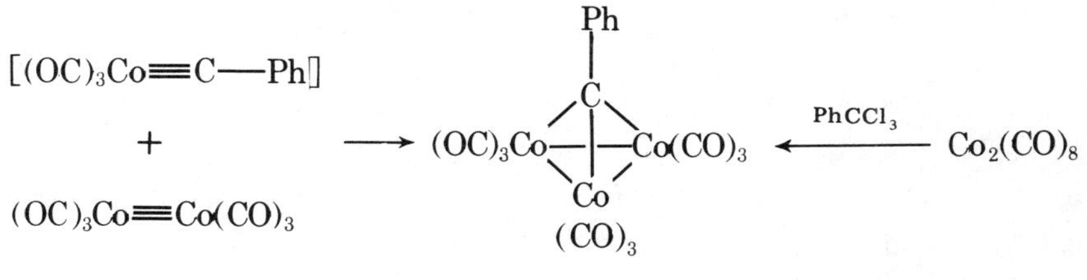

A metal-carbon double bond formally exists in metal-carbene complexes

$$\left(L_nM=C\diagup_{R'}^{R} \right)$$

These metal-carbon double bonds are generally polarized because of the difference in the electronegativities between M and C. Furthermore, the poor $d\pi$-$p\pi$ overlap between the relevant π orbitals of M and C makes the bond highly reactive. For example, organosilicon compounds containing a Si=C bond are exceedingly reactive. The formation of highly reactive silaethylenes ($R_2Si=CR_2'$) by pyrolysis of suitable silacycles has been proposed. The silaethylenes were trapped by reaction with CH_3OH or with dienes. The silaethylenes seem to dimerize very readily, giving disilacyclobutanes, and the reaction has been formally regarded as a (2 + 2) cycloaddition.[72,73]

$$[Me_2Si=CH_2] \xrightarrow{CH_3OH} Me_2Si-CH_3$$
$$\qquad\qquad\qquad\qquad\quad |$$
$$\qquad\qquad\qquad\qquad\quad OCH_3$$

$$\xrightarrow{dimerization} \begin{array}{c} Me_2Si-CH_2 \\ |\qquad\quad| \\ H_2C-SiMe_2 \end{array}$$

Other transition metal dialkylcarbene complexes are also reactive and readily give carbene dimers, probably through a dimetallocyclobutane intermediate.[74] For example, diazoalkanes (R_2CN_2) or *gem*-dihaloalkanes (R_2CX_2) react with

$$Ph_2CN_2 + [RhCl(CO)_2]_2 \xrightarrow{N_2^+} [Ph_2C=RhCl(CO)]$$

$$\downarrow dimerization$$

[structure: Ph₂C=CPh₂ ← thermal decomposition ← dirhodium bridged complex with Ph₂C, Cl, OC, CO ligands]

some low-valent metal complexes yielding ethylene derivatives ($R_2C=CR_2$). When the self-associative reaction (dimerization) of reactive carbene complexes has been hindered by the bulkiness of the auxiliary ligands or by electronic stabilization, the dialkyl- or diarylcarbene-metal complexes have been isolated and utilized for many other synthetic purposes. Thus diphenylcarbene(pentacarbonyl)tungsten has been isolated and allowed to react with olefins to yield a mixture of cyclopropane derivatives and olefin disproportionation products via a formal (2 + 2) cycloaddition between M=C and C=C bonds.[75]

$$(OC)_5W{=}C\begin{smallmatrix}Ph\\Ph\end{smallmatrix} + \begin{smallmatrix}CH_3\\CH_3\end{smallmatrix}C{=}CH_2 \xrightarrow{-CO} \left[\begin{smallmatrix}(OC)_4W{-}C(Ph)(Ph)\\|\quad\quad\quad|\\CH_3{-}C(CH_3){-}CH_2\end{smallmatrix}\right]$$

$$\left[\begin{smallmatrix}W(CO)_5\\\|\\C\\/\ \ \backslash\\CH_3\ \ CH_3\end{smallmatrix}\right] + \begin{smallmatrix}CPh_2\\\|\\CH_2\end{smallmatrix} \longleftarrow \quad \Big\downarrow +CO$$

Not isolated

$$\begin{smallmatrix}H\\ \ \\H\end{smallmatrix}\!\!\diagdown\!\!\begin{smallmatrix}CPh_2\\/\ \backslash\\C{-}C\\|\\CH_3\end{smallmatrix}\!\!\diagup\!\!CH_3$$

Although a pure (2 + 2) reaction is restricted by orbital symmetry, the present the reaction of reactive dialkylcarbene complexes with olefins is influenced by many unknown factors, and the elucidation of the nature of this important elementary reaction will contribute to the understanding of catalytic cyclopropanation and olefin metathesis.[75]

Cycloaddition involving metal-carbon unsaturation is not confined to (2 + 2) reactions. Recently a (4 + 2) reaction has been found in compounds with a Si=C, Si=Si, or Ge=C linkage.[76]

$$\begin{smallmatrix}Me\\ \ \ \diagdown\\ \ \ \ \ M{=}CH_2\\ \ \ \diagup\\Me\end{smallmatrix}$$

+

$$\begin{smallmatrix}H_2C\ \ \ \ \ \ \ CH_2\\ \ \diagdown\ \ \ \diagup\\ \ \ \ C{-}C\\ \ \ \ |\ \ \ |\\ \ \ \ H\ \ H\end{smallmatrix}$$

M = Si, Ge

$$\longrightarrow \begin{smallmatrix}Me\\ \ \diagdown\\Me{-}M\\ \ \ \ \ \diagup\ \diagdown\end{smallmatrix}$$

$$Me_2Si{=}SiMe_2$$
+
$$\begin{smallmatrix}H_2C\ \ \ \ \ \ \ CH_2\\ \ \diagdown\ \ \ \diagup\\ \ \ \ C{-}C\\ \ \ \ |\ \ \ |\\ \ \ \ H\ \ H\end{smallmatrix}$$

$$\longrightarrow Me_2Si{-}SiMe_2$$

These reactive M=C or Si=Si bonds have been usually generated by pyrolysis (cycloreversion) of the relevant metallocycles.

The nonbonding metal electron pair has been usually regarded as the two-electron component of the cycloaddition, and the formation of *cis*-dihydrides by the oxidative addition of a hydrogen molecule to a low-valent metal [e.g., $RhCl(PPh_3)_3$] can be a (2 + 2) process that is allowed by participation of metal *d* orbitals (as described below). The formation of an olefin complex also belongs to the same category. Here, the olefin complex has been regarded as a metallocyclopropane.

$$M:^{*)} + \begin{matrix} H \\ | \\ H \end{matrix} \rightleftharpoons M\begin{matrix} H \\ \\ H \end{matrix}$$

$$M:^{*)} + \begin{matrix} C \\ \| \\ C \end{matrix} \rightleftharpoons M\begin{matrix} C \\ \\ C \end{matrix}$$

Orbital symmetry rules predict these reactions to be thermally forbidden. Actually, no immediate reaction occurs between hydrogen and AsR_3 or SeR_2, which contain as the metal atom *nontransition elements* with a nonbonding electron pair consisting of *s* and *p* electrons. The addition of hydrogen or an olefin

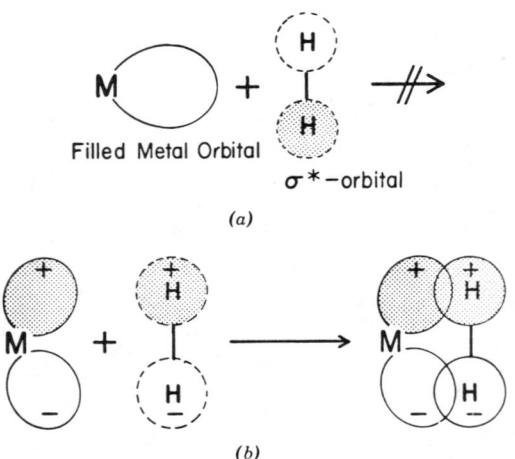

Fig. 38. Concerted addition of H_2 to (*a*) metals of nontransition elements and (*b*) transition metals, showing (*a*) a side wise interaction between a filled metal σ orbital and a vacant $H_2\sigma^*$ orbital and (*b*) a side wise interaction between a filled metal *d* orbital and a vacant $H_2\sigma^*$ orbital.

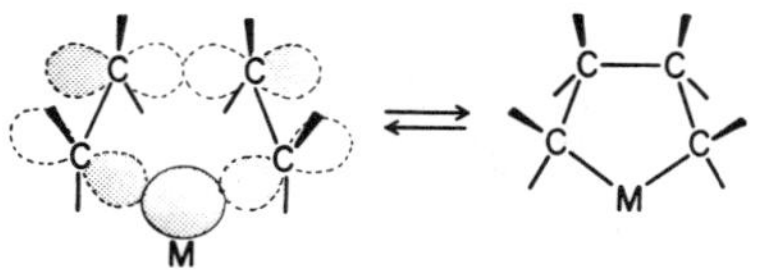

Fig. 39. A symmetry-allowed termolecular electrocyclic reaction forming a Metallocyclopentane ring.

to form a complex is a reaction that frequently occurs with low-valent *transition metal* complexes, however. The symmetry of the *d* orbitals has been thought to play a decisive role in these reactions by lowering the energy barrier imposed by orbital symmetry rules (see Fig. 38).

Formation of metallocycles from a metal atom and two or more unsaturated bonds (metallocyclization) has also been predicted by symmetry rules. Thus catalytic cyclooligomerization of olefins or acetylenes involves such metallocycles as key intermediates. A simple case has been the concerted formation of metallocyclopentane from two alkene molecules and a metal complex. If the metal can supply two electrons, the metallocyclization will be a thermally allowed process (see below).[77] Some examples of this type of metallocyclization are known (Fig. 39), but there is no evidence for a concerted reaction involving a trimolecular transition state. An analogous cycloaddition between a rhodium complex and diacetylene may occur in a concerted fashion if the two C≡C bonds are sterically well situated to promote the following reaction.

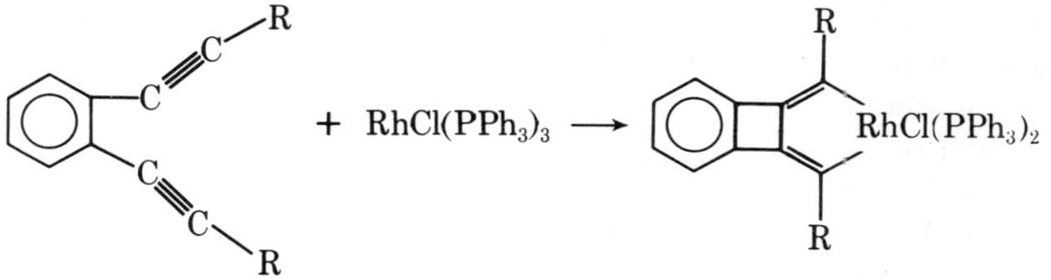

$$\text{(diyne)} + RhCl(PPh_3)_3 \longrightarrow \text{(cyclobutadiene complex)}\, RhCl(PPh_3)_2$$

Cyclooligomerization of conjugated dienes with zerovalent nickel-ligand catalysts may also involve cycloaddition and electrocyclic reactions as the most important catalytic steps. Exclusive formation of substituted *cis*-divinylcyclobutane from two molecules of *trans*-pentadiene is explained by the scheme in Fig. 40.[78]

Here zerovalent nickel has been assumed to supply two electrons to the cyclization process $(4 + 4 + {}_M2)$, where the participating metal electrons have been expressed as ${}_M2$. In summary, stereoelectronic effects generally observed in many cycloadditions between organic unsaturations seem also to play an

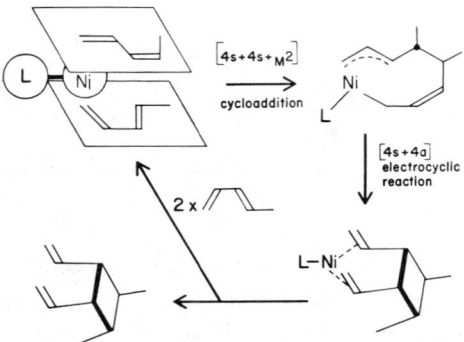

Fig. 40. A proposed mechanism of the stereospecific cyclodimerization of *cis*- and *trans*- pentadiene.

important role in influencing the stereoselectivity of some homogeneous catalyses.

2.5. Fluxionality and Polytopal Rearrangements

In contrast to the well-known rigidity of the tetrahedral valence angle of saturated carbon, the stereochemical disposition of ligands in some transition metal complexes is variable in solution even at −80°. The variation of the shape of metal complexes without breaking any metal-ligand bonds occurs mostly by changes in the interligand angles. When such changes occur in solution at ambient temperature, the complexes are called "stereochemically nonrigid." These stereochemically nonrigid complexes are important because they are frequently catalytically active, and the activity has sometimes been associated with the variable M—L bond angle (see Figs. 23 to 25 and accompanying text).

Stereochemically nonrigid molecules change their structures by "fluxionality" or "polytopal rearrangements." The former process occurs among chemically equivalent structures, and the difference in many cases has been discernible only by the isotopic substitution at the ligand as shown in Fig. 41.

Polytopal rearrangement refers to the process by which polyhedral coordination spheres such as octahedral or prismatic are interconverted. Typical examples of polytopal rearrangements (or polytopal changes) are shown in Figure 42.[79]

An important feature of these stereochemical processes is that M—L bonds are never broken. Structural changes of metal complexes through dissociation

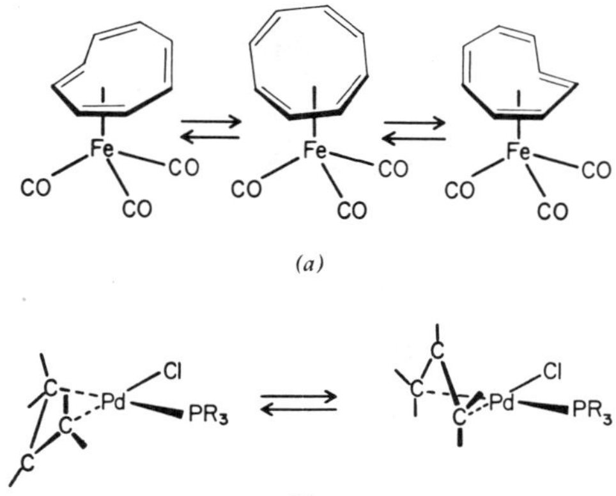

Fig. 41. Examples of fluxional organometallic molecules. (a) Ring whizzing: tricarbonyl(η^4-cyclooctatetraene)iron(0). (b) A flipping η^3-allyl ligand: chloro(η^3-allyl)(*tert*-phosphine)palladium(II).

or association of ligands belong neither to fluxionality nor to polytopal rearrangements.

The fluxional behavior of a complex has been usually detected by its nmr spectrum. For example, chemical shift values of protons or carbons of the ligands have been recorded while the temperature of a solution of the complex in question has been varied. Typical nmr results are illustrated in Figure 43 for $Fe(CO)_3(\eta^4-C_8H_8)$, which was one of the earliest studied examples of a fluxional

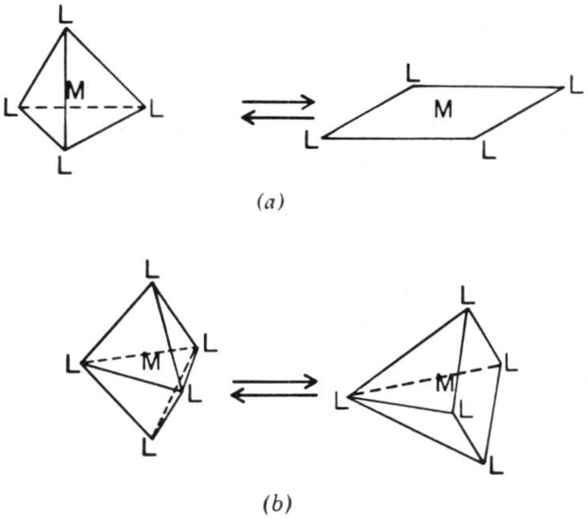

Fig. 42. Examples of polytopal rearrangements. (a) Four-atom family (tetrahedral ⇌ square planar). (b) Five-atom family (trigonal bipyramidal ⇌ square pyramidal).

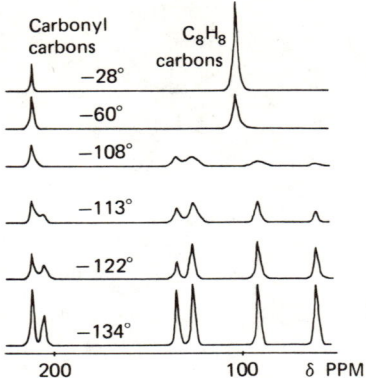

Fig. 43. The temperature-dependent ^{13}C nmr spectrum of $Fe(CO)_3$ (η^4-C_8H_8).

molecule.[80] Inspection of Fig. 43 reveals that one sharp signal splits into four nonequivalent signals on lowering the temperature. The sharp ^{13}C nmr signal as well as a sharp ^1H nmr singlet observed above $-60°$ indicates equivalence of the C_8 ring carbon atoms on the nmr time scale (i.e., $\sim 10^{-2}$ sec). The static structure of the molecule is of the (1-4)-tetrahapto type, and four carbons out of eight are bound to the $Fe(CO)_3$ group as evidenced by an x-ray diffraction analysis and by a solution ir spectroscopic study. The activation energies for the fluxionalization of $Fe(CO)_3\eta^4$-C_8H_8 and $Ru(CO)_3\eta^4$-C_8H_8 are 8.1 and 8.6 kcal, respectively (cf. Cotton and Hunter[80]). Detailed analysis of the nmr line shape has revealed a 1,2-shift mechanism for $M(CO)_3$-η^4-C_8H_8.

1,2-Shift process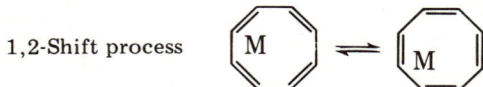

Polytopal changes have also been revealed by nmr spectra. The ^1H nmr spectrum of $H_2Fe(Ph_2PCH_2CH_2PPh_2)_2$ shows a symmetrical quintet (1:4:6:4:1) peak in the hydride region at room temperature.[81] The quintet arises from coupling of the two equivalent hydride protons with four equivalent ^{31}P nuclei. However the x-ray structure (Fig. 44) of $H_2Fe(Ph_2PCH_2CH_2PPh_2)_2$ indicates the nonequivalence of the P atoms. Rapid fluxionalization must be taking place in solution with mutual positions of H and P ligands rapidly changing to result in a time-averaged equivalence of all the P ligands.

The polytopal rearrangement pathways shown in Fig. 45 have been proposed on the basis of the line shape analysis of nmr peaks.[81] The possible consequence of these dynamic intramolecular processes toward catalysis remains to be evaluated.

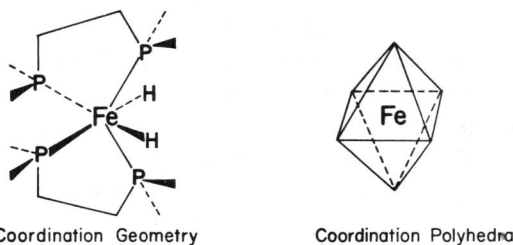

Fig. 44. The distorted octahedral (C_2 symmetry) geometry of a nonrigid molecule, $H_2Fe(Ph_2PCH_2CH_2PPh_2)_2$.

Some polyhydrido complexes [e.g., $IrH_5(PR_3)_2$] have been found to be very active catalysts for the deuteration of aromatic compounds with D_2. Metallocene hydrides also exhibit fluxional or polytopal behavior. The equivalence of hydride ligands in $[MoH_3Cp_2]^+$ has been evidenced by the single hydride nmr peak at

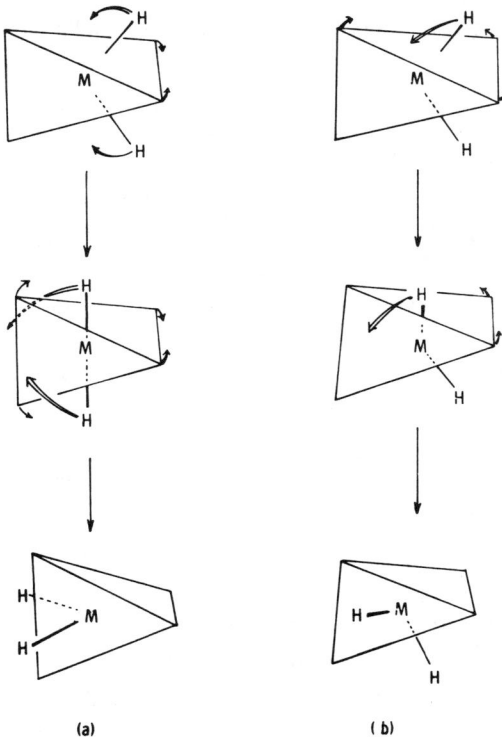

Fig. 45. Possible mechanisms of polytopal rearrangements of $MH_2(PR_3)_4$. (a) D_{2d} Path (*trans*-dihydride path). (b) Tetrahedral jump.

room temperature. The dynamic process described above is shown in Fig. 46. It has also been suggested that the corresponding dihydride MoH$_2$Cp$_2$ is in a thermal dynamic equilibrium (see Fig. 46).[82]

The parallel metallocene structure seems to be responsible for the reactivity toward π-acidic olefins because it has a π-basic metal site ready to bind olefins and leads to insertion into the Mo—H bond (see Section 2.3).

2.6. Reactions of Coordinated Ligands

When a small molecule is coordinated to a metal, its electronic state as well as its geometry changes depending on the perturbation due to the metal. Some of these changes have been successfully utilized in many homogeneous catalyses. For example, the catalytic hydration of an acetylenic triple bond occurs using aqueous Hg^{2+} ion. The high polarizing ability of Hg^{2+} causes the acetylenic carbons to become partially positive on η^2-complexation with Hg^{2+} (formation of mercurinium ion) and to become susceptible to nucleophilic attack by, for example, MeOH or H$_2$O.

Similar but weaker polarization occurs with many other cations of metals (e.g., Pt^{2+}, Pd^{2+}, Cu$^+$, or Zn^{2+}). Zinc salts are effective in the catalytic vinylation of alcohols by acetylene at high temperature and pressure. This process is known as "Reppe's vinylation reaction." [83] In the following probable mechanism involving nucleophilic attack of the alcoholate anion, η^1-coordination may be important (see Chapter Section 2.7):

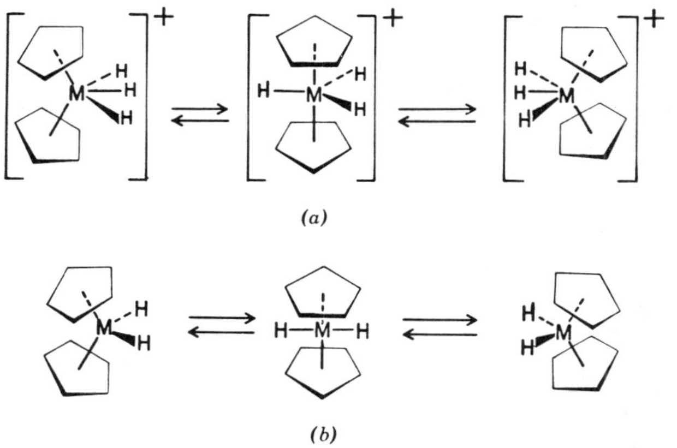

Fig. 46. Polytopal rearrangements. (*a*) [MH$_3$Cp$_2$].$^+$ (*b*) MH$_2$Cp$_2$.

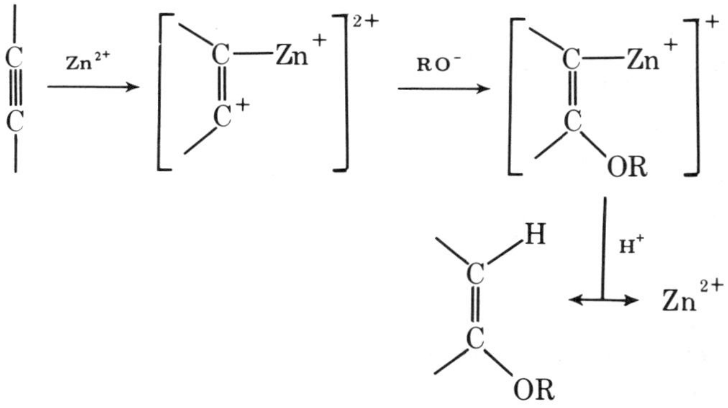

Olefins coordinated to PdCl$_2$ species are also activated toward nucleophilic attack by H$_2$O or AcO$^-$. Ethylene has thus been converted to acetaldehyde or vinylacetate by stoichiometric reaction with PdCl$_2$/aq HCl or Pd(OAc)$_2$/AcOH systems, respectively.[84]

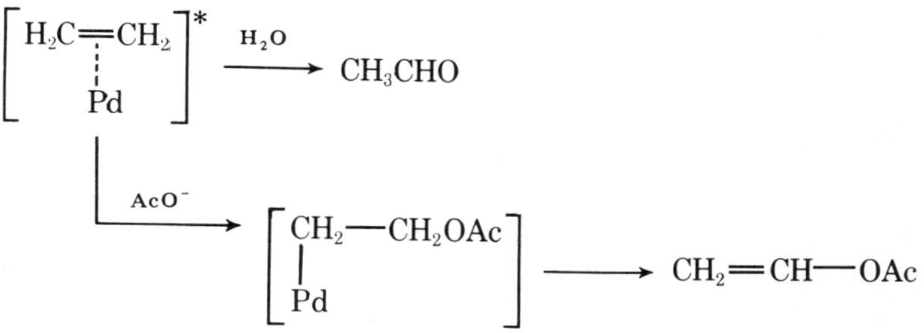

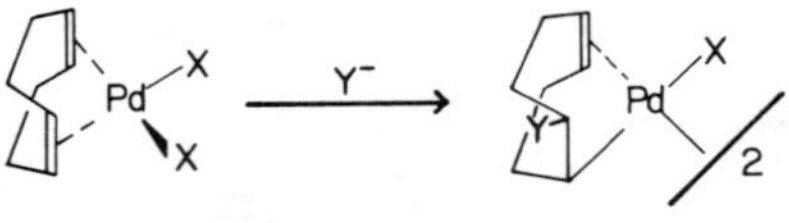

Scheme 7

In Scheme 7 a chelating diene coordinates to a Pd²⁺ compound, and the resulting intermediate σ-alkyl–palladium complex has been isolated. The stereochemistry of the nucleophilic attack has been found to be trans to the metal.[85]

Hydration of organic nitriles to the amides also proceeds readily when the nitrogen end of the nitrile coordinates to an electropositive metal.[86] The nucleophilic attack of water at the unsaturated carbon leads to the formation of the amide through a hydroxyimide (see Chapter 5, Section 1 on Cu^{2+}-catalyzed hydration of $CH_2{=}CH{-}CN$).

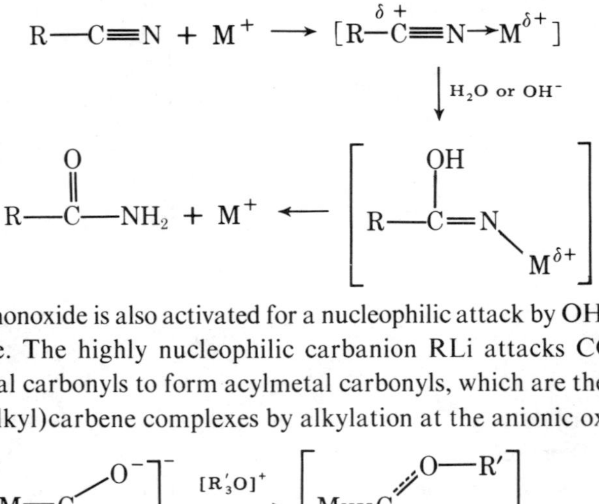

$$R{-}C{\equiv}N + M^+ \longrightarrow [R{-}\overset{\delta+}{C}{\equiv}N{\rightarrow}M^{\delta+}]$$

$$\downarrow H_2O \text{ or } OH^-$$

$$R{-}\underset{\underset{O}{\parallel}}{C}{-}NH_2 + M^+ \longleftarrow \left[R{-}\underset{\underset{OH}{|}}{C}{=}N{\diagdown}_{M^{\delta+}} \right]$$

Carbon monoxide is also activated for a nucleophilic attack by OH^- or amines, for example. The highly nucleophilic carbanion RLi attacks CO groups of various metal carbonyls to form acylmetal carbonyls, which are then converted to alkoxy(alkyl)carbene complexes by alkylation at the anionic oxygen end.[87]

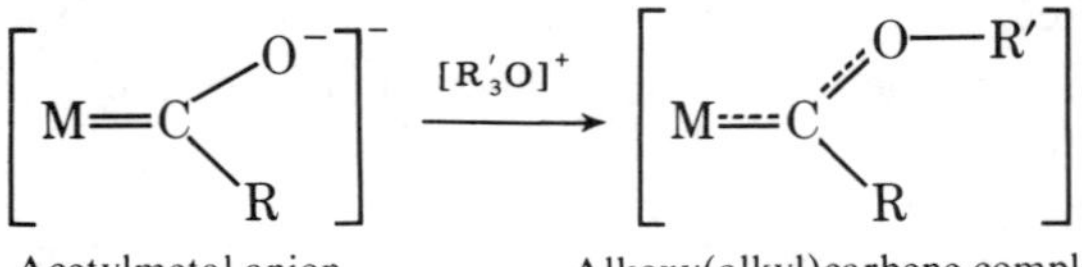

$$\left[M{=}C\diagup^{O^-}_{\diagdown R} \right] \xrightarrow{[R'_3O]^+} \left[M{\doteq}C\diagup^{O{-}R'}_{\diagdown R} \right]$$

Acetylmetal anion   Alkoxy(alkyl)carbene complex

$$M{-}C{\equiv}O \xrightarrow{OH^-} \left[M{-}C\diagup^{OH}_{\diagdown O} \right] \longrightarrow CO_2 + [M{-}H]^-$$

$$\downarrow RNH_2, -H^+$$

$$\left[M{-}C\diagup^{NHR}_{\diagdown O} \right] \xrightarrow{RNH_2} (RNH)_2CO$$

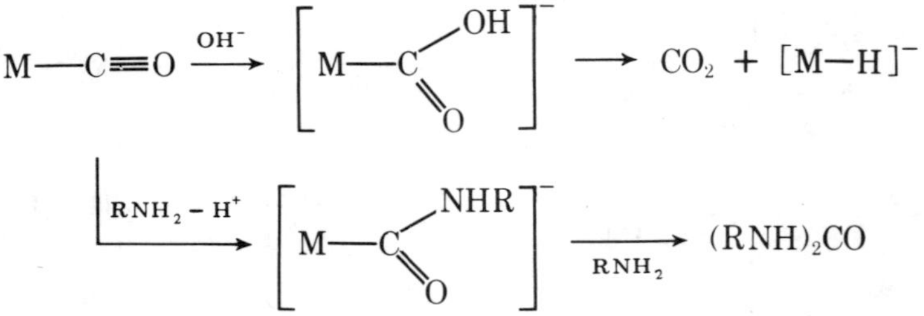

Isonitriles behave similarly to carbon monoxide when coordinated to an electropositive metal. Reaction of some isonitrile complexes with alcohols or amines yields carbene complexes. Generally these alkoxy(amino) or diamino-carbene platinum or palladium complexes are thermally stable.

$$PtCl_2(CN-R)_2 \xrightarrow{CH_3OH} PtCl_2\left(C\begin{array}{c}OCH_3\\NR\\|\\H\end{array}\right)_2$$

Hydrogen peroxide has been activated for olefin epoxidation by catalysis using W(VI) or Mo(VI) complexes. The highly positive character of the metal induces strong polarization in the O—O bond. The positively charged oxygen in M—O—O$^+$ has been postulated as the catalytically active species. The positively charged oxygen has a high affinity for the electronic cloud of an olefin π-bonding orbital and will be readily transferred to give epoxides.[88]

$$H_2O_2 \xrightarrow[+L]{WO_4^{2-}} \left[\begin{array}{c}\text{structure}\end{array}\right]$$

L = HMPA

An oxygen molecule end-coordinated to Co(II) chelates has been thought to be activated for some reactions. The esr spectrum of O_2 Co(bzacen)/(pyridine) has been interpreted to indicate the presence of O_2^-, which is reactive toward organic radical scavengers. Because of the possible relationship to biochemical enzymatic oxygenation, this coordination effect of some transition metals is important.[89] (see p. 148).

Molecular nitrogen has also been made basic enough to be protonated at the terminal atom when coordinated to some basic transition metals in an end-on manner. Thus one of the coordinated nitrogen molecules of $(N_2)_2W(PMe_2Ph)_4$ has been converted to ammonia by protonation followed by reduction (see Chapter 5, Section 13).[90] The intermediate protonated species shown below has been isolated when a suitable chelating diphosphine was used as an auxiliary

$$\textit{cis-}(N_2)_2W(PMe_2Ph)_4 \xrightarrow{\text{aq } H_2SO_4} NH_3 + N_2 + W(VI) \text{ species}$$

ligand. Some of the coordinated nitrogen in $(C_5Me_5)_2Zr(N)_2$—N≡N—$Zr(N_2)(C_5Me_5)_2$ has been also reduced to ammonia and hydrazine by protonation. The protolytic reductive cleavage of N≡N bond forms the basis for titanium catalyzed nitrogen fixation with sodium naphthalenide.[91]

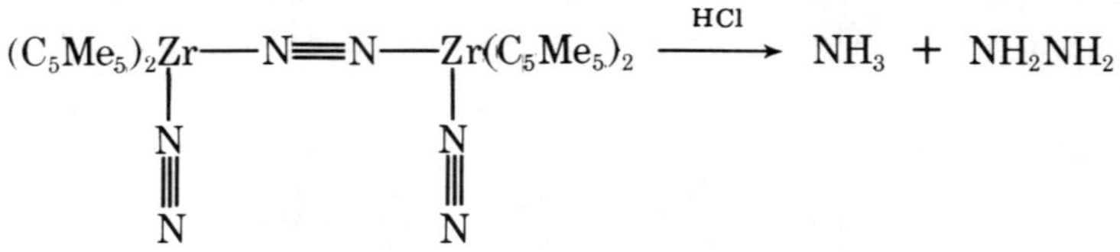

$$(C_5Me_5)_2Zr\text{—}N\!\!\equiv\!\!N\text{—}Zr(C_5Me_5)_2 \xrightarrow{HCl} NH_3 + NH_2NH_2$$
$$\overset{|}{\underset{\overset{\|}{N}}{N}} \qquad \overset{|}{\underset{\overset{\|}{N}}{N}}$$

HCN (or CNH) has been selectively *N*-alkylated with some branched olefins when Cu^I salts are present. This was caused by a strong σ-type coordination of

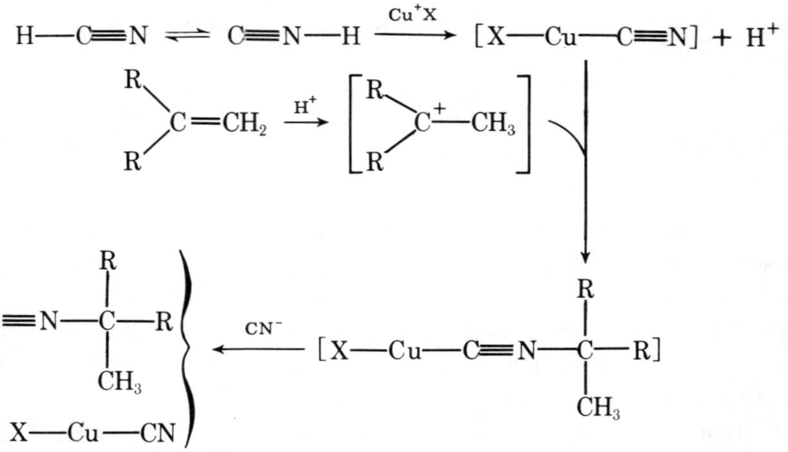

$$H\text{—}C\!\!\equiv\!\!N \rightleftharpoons C\!\!\equiv\!\!N\text{—}H \xrightarrow{Cu^+X} [X\text{—}Cu\text{—}C\!\!\equiv\!\!N] + H^+$$

CNH to Cu(I) followed by an attack of a *tert*-carbonium ion on the nitrogen end. The alkylated CN ligand (i.e., coordinated isonitrile) was easily freed from the metal by addition of KCN.[92]

2.7. σ-π Rearrangements

Since the isolation of triphenyltris(tetrahydrofuran)chromium(III) as an intermediate in the synthesis of Hein's bis(η^6-arene)chromium complex, numerous examples of an important class of organometallic reactions, σ-π rearrangements, have been reported.[93] These rearrangements are of both theoretical and practical interest, since they appear to be very important in homogeneous catalytic processes, and a knowledge of the factors that govern this phenomenon is essential. A σ-π rearrangement is an intramolecular reaction in which an organic group (e.g., carbon ligand) σ-bonded (η^1) to a metal becomes π-bonded (η^n) to the metal. The reverse of this process, π-σ ($\eta^n \rightarrow \eta^1$), also occurs.

An early example of the formation of bis(η^6-arene)chromium complexes by the thermal decomposition of poly(η^1-aryl)chromium compounds was discovered by F. Hein and co-workers. The reactions are now known to proceed by σ-π rearrangement of a labile η^1-arylmetal moiety, illustrated in Scheme 8.[94] Among various kinds of rearrangement, interconversion between η^1-(σ) and η^2-(π)-ligands is the most fundamental one.

For example, η^2-carbon ligands (i.e., π-olefin ligands) have been transformed to the corresponding η^1-ligands by thermal or other excitations as shown in Scheme 9. Since the η^1-species are in more or less excited states, they have partially ionic and partially radical character. The classical representation of

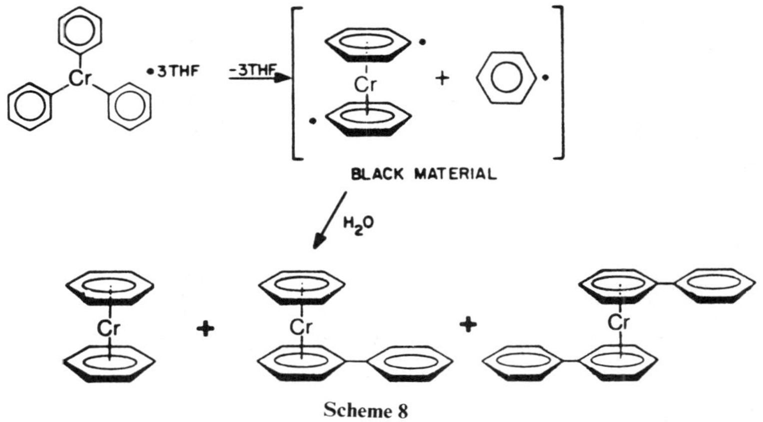

Scheme 8

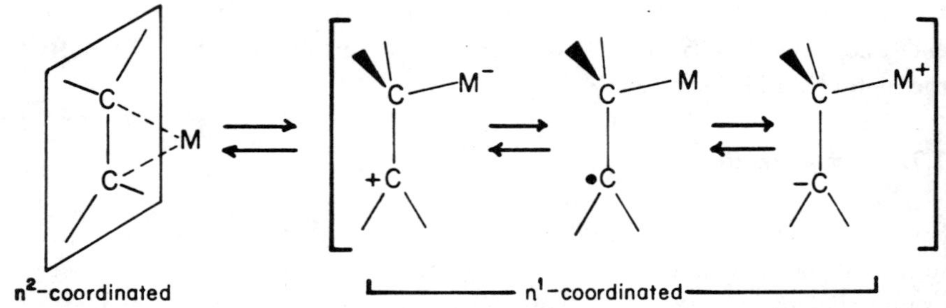

[Scheme 9 figure: η²-coordinated ⇌ η¹-coordinated forms]

Scheme 9

these species has put +, −, or · signs on them as a formalism. The correct representation is a partial intermediate between all these species. Unless stabilized by combination with suitably reactive partners (anionic, cationic, or radical reagents), the η^1-species remain excited. Usually the formation of stabilized products having σ-(η^1)-bonded ligands has been cited as evidence for the σ-π rearrangement.

The reactive η^1-species generated from η^2-olefin ligands have frequently been assumed as intermediates in various reactions (Scheme 10).

Most σ-π rearrangement reactions result in reaction at the metal center or reaction on the ligand itself. Dissociation of an auxiliary ligand (L′) (see below) may result in a σ-π rearrangement by increasing hapto numbers of polyhapto ligands; addition of the ligand to the metal causes the reverse rearrangement. The loss of an auxiliary ligand leaves a metal coordinatively unsaturated, and the σ-π rearrangement of ligand A will satisfy the coordination requirement by participation of more than one electron pair from the σ-π ligand.

$$M(\eta^1-A)L'_n \rightleftharpoons M(\eta^1-A)L'_{n-1} + L' \underset{\sigma\text{-}\pi \text{ rearrangement}}{\rightleftharpoons} M(\eta^m-A)L'_{n-1} + L'$$

For example, when σ-allyl (η^5-cyclopentadienyl)tricarbonylmolybdenum(II) and tungsten(II) have been irradiated with ultraviolet light, a σ-π rearrangement occurs yielding the π-allyl or η^3-allyl complex. On addition of carbon monoxide, the reverse π-σ rearrangement has been observed.[97]

$$CpM(CO)_3(\eta^1\text{-}CH_2-CH=CH_2) \underset{}{\overset{h\nu,-CO}{\rightleftharpoons}} CpM(CO)_2(\eta^3\text{-}CH_2-CH-CH_2)$$

Photochemical cis-trans isomerization of η^2-coordinated olefin has also been proposed to involve a photochemically activated η^1-olefin complex as an important intermediate.[98]

$$HC\equiv CH + ROH \xrightarrow{Zn^{II}} RO-CH=CH_2$$

Mechanism:

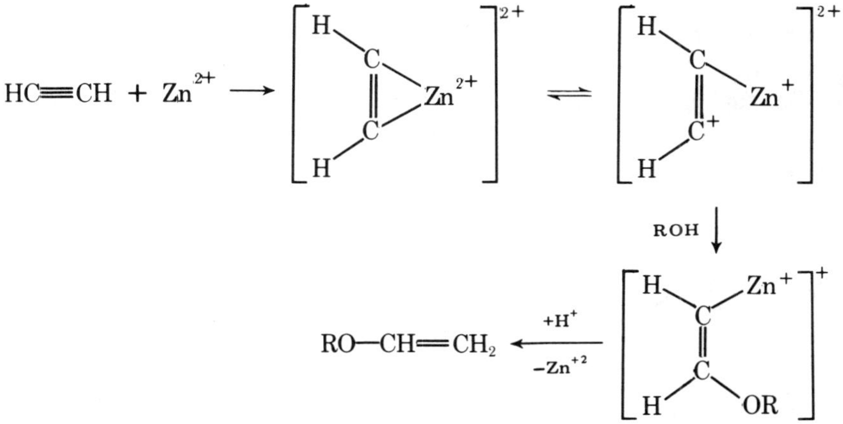

$$HC\equiv CH + Zn^{2+} \longrightarrow \left[\begin{array}{c}H\\ \diagdown C\\ \parallel \quad Zn^{2+}\\ C\\ \diagup\\ H\end{array}\right]^{2+} \rightleftharpoons \left[\begin{array}{c}H\\ \diagdown C\\ \parallel \quad Zn^{+}\\ C^{+}\\ \diagup\\ H\end{array}\right]^{2+}$$

$$RO-CH=CH_2 \xleftarrow[-Zn^{+2}]{+H^+} \left[\begin{array}{c}H \quad Zn^{+}\\ \diagdown \diagup\\ C\\ \parallel\\ C\\ \diagup \diagdown\\ H \quad OR\end{array}\right]^{+}$$

$\downarrow$ ROH

(a)

$$\begin{array}{c}CH_2-CH_2\\ |\quad\quad |\\ OH\quad OH\end{array} \longrightarrow CH_3CHO$$

$$\begin{array}{c}CH_2-CH_2\\ |\quad\quad |\\ NH_2\quad OH\end{array}$$

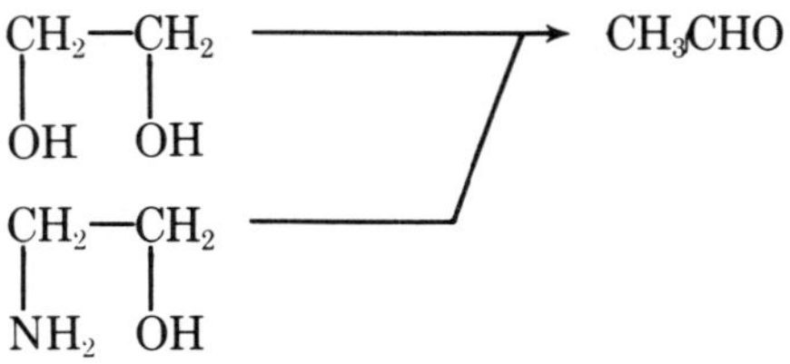

Mechanism:

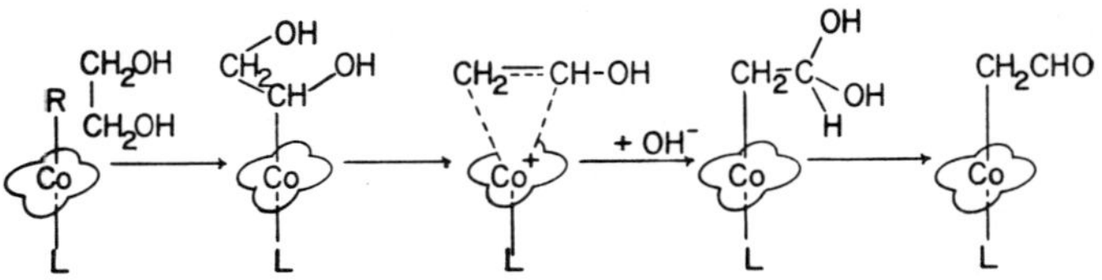

(b)

Scheme 10. (a) Addition of nucleophiles to an (η^2-acetylene) Zn(II) complex.[95] (b) Rearrangements catalyzed by coenzyme B_{12}-dependent enzymes.[96]

$$\text{H}\underset{\text{D}}{\overset{\text{M}}{\underset{|}{\text{C}}\text{—}\underset{|}{\text{C}}}}\text{H} \quad \underset{}{\overset{h\nu}{\rightleftarrows}} \quad \text{complex intermediate} \quad \underset{}{\overset{h\nu}{\rightleftarrows}} \quad \text{H}\underset{\text{D}}{\overset{\text{M}}{\underset{|}{\text{C}}\text{—}\underset{|}{\text{C}}}}\text{D}$$

$$M = W(CO)_5$$

Creating a reactive center on the η^1- or η^2-ligand can also cause a σ-π rearrangement. β-Hydride abstraction from the σ-alkyl ligand and protonation at the η^2-olefin ligand are perhaps the most well-known examples.

Some transition metal σ-alkyl complexes with the alkyl group being ethyl, n-propyl, or isopropyl react with the triphenylmethyl cation to form olefinic complexes.[99]

$$CpFe(CO)_2\text{—}CH(CH_3)_2 \xrightarrow{Ph_3C^+} [Fe\text{—}CH(CH_2)(CH_3)]^+ \longrightarrow$$

$$[CpFe(CO)_2(CH_2\!=\!CH)]^+ \atop |\atop CH_3$$

Other transition metal σ-alkyl complexes, such as cyanoalkyl complexes in which the alkyl moiety has an unsaturated group, undergo reversible protonation to form olefinic complexes.[100]

$$CpFe(CO)_2(\text{—}CH_2CN) \underset{}{\overset{H^+}{\rightleftharpoons}} [CpFe(CO)_2(CH_2\!=\!C\!=\!NH)]^+$$

Not all σ-π rearrangements require chemical initiation. Some metal complexes exist in a dynamic equilibrium between the σ and π forms. For example, a reversible σ-π rearrangement has been postulated to account for observed regioselective insertion into metal–η^2-acetylene bonds.[101]

$$\left[M\overset{R^1}{\underset{R^2}{\overset{|}{\underset{|}{C}}\!\!\cdots\!\!\underset{|}{C}}} \rightleftharpoons M\overset{R^1}{\underset{}{\overset{|}{C}}}\text{=}\underset{R^2}{\overset{|}{C}} \right]^+ \xrightarrow{R^3\text{—}C\!\equiv\!C\text{—}R^4} M\overset{R^1}{\underset{}{\overset{|}{C}}}\text{=}\underset{R^3}{\overset{R^2}{\underset{}{\overset{|}{C}}}}\text{—}\underset{R^4}{\overset{}{C}}$$

η^2-Acetylene complex $\quad\eta^1$-Acetylene complex

Thermal cis-trans isomerization of some olefin complexes with electronegative substituents can also be cited as an example.[102]

Nmr spectroscopy has shown that the allyl ligand in some transition metal η^3-allyl complexes at room temperature has all the terminal hydrogens magnetically equivalent.[103] In most cases this equivalence results from σ-π rearrangements ($\eta^1 \rightarrow \eta^3$) that are rapid on the nmr time scale (ca. 30 times/sec).

S = coordinating solvent

Let us examine the differences in bonding in typical σ and π complexes of carbon ligands. A σ complex (η^1) involves a metal ligand bond that results mainly from the interaction of a metal σ orbital with one of the carbon sp^n hybrid orbitals. A π complex involves a metal-ligand bond consisting of metal orbitals and ligand-carbon orbitals extending over two or more centers; therefore it is composed of two components—a σ bond between a metal acceptor orbital and a ligand molecular orbital of σ symmetry, and a π bond between a metal d orbital or a $p\pi$-$d\pi$ hybrid and the ligand π^* (antibonding) molecular orbital in which electron density flows from the metal to the ligand. The stabilities of the two bonding modes are, of course, primarily dependent on the electronic properties of the metal and of the auxiliary ligands.

Since the two orbital interactions are operating to different extents in these transition metal η^2-olefin complexes, the factors favoring η^2 coordination are not easily discerned. When the π interaction (back-donation) is considerable, metals favoring back-donation (e.g., metals with high metal-electron density) prefer the η^2 coordination. However if the σ interaction predominates in metal-ligand bonding, the preference for η^2-coordination will be weaker.

The relative stability of the two species, being dependent on the polarization of the metal center, has also been correlated to the concept of "hard" and "soft" acids and bases. The small relatively nonpolarized σ orbitals of organic groups are harder than the more delocalized electron cloud of π orbitals. Thus "hardening" the metal by placing electron-withdrawing groups on it will favor the σ complex. Conversely, "softening" the metal will favor the π complex.

Recently much work has been done in an attempt to better define the role of σ-π rearrangements in catalytic processes. For example, it may be a key step in the Wacker acetaldehyde process. Kinetic and isotope effect data seem to indicate that the initially formed η^2-ethylenepalladium complex is converted

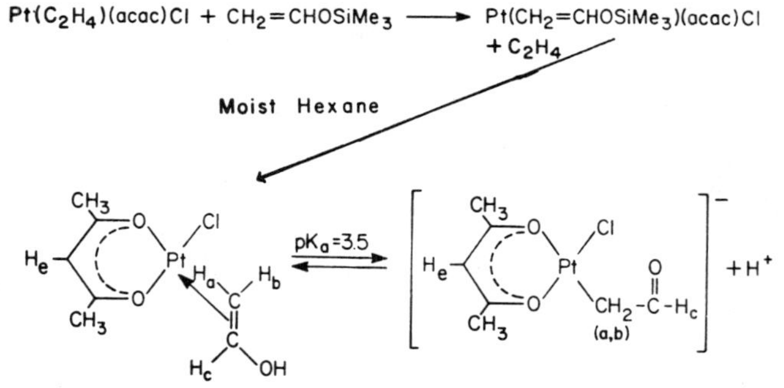

Scheme 11

to a η^1-β-hydroxyethylpalladium species in the kinetically important catalytic step. Further transformation of this σ-alkyl species has been proposed to proceed either directly to acetaldehyde or through η^2-vinylalcohol-palladium species to acetaldehyde.[85,86]

Here facile and rapid conversion of the η^2-vinylalcohol to η^1-formylmethyl (or β-oxoethyl) has been assumed. The postulated σ-π rearrangement has been confirmed by the preparation and by structure determination of a (η^2-vinylalcohol)platinum complex as shown in Scheme 11.[104]

The bond length between the platinum and the terminal olefinic carbon (2.098 Å) is significantly shorter than that between the platinum and the internal carbon (2.222 Å). Considering the ease of σ-π rearrangement by protonation, the η^2-bonded vinylalcohol ligand is approaching the corresponding σ structure even in its ground state. The ^{1}H nmr spectrum in polar solvents shows an A$_2$X pattern for the vinyl protons rather than the expected ABX pattern such as that found for the η^2-vinylsilylether structure.

SELECTED READINGS

General

F. Basolo and R. G. Pearson, *Mechanisms of Inorganic Reactions*, 2nd ed., Wiley, New York, 1967.

J. March, *Advanced Organic Chemistry*, McGraw-Hill, New York, 1968.

N. Sutin, in *Inorganic Biochemistry*, G. L. Eichhorn, Ed., Elsevier, Amsterdam, 1973, p. 611.

F. A. Cotton and G. Wilkinson, *Advanced Inorganic Chemistry*, 4th ed., Wiley, New York, 1980.

Catalytic Oxidations

R. A. Sheldon and J. K. Kochi, *Adv. Catal.*, 25, 272 (1976).

Oxidative Additions

J. K. Stille, K. S. Y. Lau, and P. K. Wong., *J. Am. Chem. Soc.*, 98, 5832 (1976).

J. A. Osborn, in *Organotransition Metal Chemistry*, Y. Ishii and M. Tsutsui, Eds., Plenum Press, New York, 1974.

Insertions

A. Nakamura and S. Otsuka, *J. Am. Chem. Soc.*, 95, 7262 (1973).

Polytopal Rearrangements

E. L. Muetterties, *MTP International Review of Science, Inorganic Chemistry* Series, Vol. 1, Butterworths, London, 1974.

F. A. Cotton, and L. M. Jackman Eds., *Dynamic NMR Spectroscopy*, Academic Press, New York, 1975.

F. A. Cotton, *J. Organomet. Chem.*, 100, 29 (1975).

Cycloadditions

R. B. Woodward and R. Hoffmann, *The Conservation of Orbital Symmetry*, Verlag Chemie–Academic Press, New York, 1970.

G. B. Gill and M. R. Willis, *Pericyclic Reactions*, Chapman & Hall, London, 1974.

R. G. Pearson, *Acc. Chem. Res.*, 4, 152 (1971).

Charge Transfers

K. Tamaru and M. Ichikawa, *Catalysis by Electron Donor-Acceptor Complexes*, Kodansha-Halsted, Tokyo, 1975.

σ-π Rearrangements

M. Tsutsui and A. Courtney, "σ-π Rearrangements of Organotransition Metals," in *Advances in Organometallic Chemistry*, Vol. 16; Academic Press, New York, 1977, p. 241.

M. Hancock, M. N. Levy, and M. Tsutsui, *Organometallic Reactions*, Vol. 4, Wiley-Interscience, New York, 1972, p. 1.

Selected Readings

References for Chapter 4

1. D. J. Cram, *Fundamentals of Carbanion Chemistry*, Academic Press, New York, 1965, pp. 130.
2. J. March, *Advanced Organic Chemistry*, McGraw-Hill, New York, 1968, pp. 376.
3. L. M. Stock and H. C. Brown, *Adv. Phys. Org. Chem.*, 1, 35 (1963).
4. J. E. Huheey, *Inorganic Chemistry*, Harper and Row, New York, 1972, p. 225.
5. F. Basolo and R. G. Pearson, *Mechanisms of Inorganic Reactions*, 2nd ed., Wiley, New York, 1967.
6. F. Basolo, J. Chatt, H. B. Gray, R. G. Pearson, and B. L. Shaw, *J Chem. Soc.*, 1961, 2207.
7. F. Basolo, H. B. Gray, and R. G. Pearson, *J. Am. Chem. Soc.*, 82, 4200 (1960).
8. Ref. 5, pp. 387.
9. (a) J. O. Edwards and R. G. Pearson, *J. Am. Chem. Soc.*, 84, 16 (1962); (b) C. G. Swain and C. B. Scott, *J. Am. Chem. Soc.*, 75, 141 (1953).
10. Ref. 5, pp. 399.
11. G. N. Schrauzer, *Angew. Chem., Int. Ed.*, 15, 417 (1976).
12. See ref. 5, pp. 399.
13. See ref. 5, pp. 140.
14. N. Sutin, in *Inorganic Biochemistry*, G. L. Eichhorn, Ed., Elsevier, Amsterdam, 1973, pp. 611.
15. Ref. 5, pp. 475–476; ref. 14.
16. Ref. 5, pp. 466, pp. 475.
17. T. J. Meyer, *Acc. Chem. Res.*, 11, 94 (1978).
18. R. H. Holm and J. A. Ibers, in *Iron-Sulfur Proteins*, Vol. 3, W. Lovenberg, Ed., Academic Press, New York, 1977, pp. 229, 265.
19. S. Wherland and H. B. Gray, in *Biological Aspects of Inorganic Chemistry*, A. W. Addison et al., Eds., Wiley-Interscience, New York, 1977, pp. 289.
20. R. J. P. Williams, G. R. Moore, and P. E. Wright, in ref. 19, pp. 369.
21. R. A. Sheldon and J. K. Kochi, *Adv. Catal.*, 25, 272 (1976).
22. E. Collinson, F. S. Dainton, B. Mile, S. Tazuke, and D. R. Smith, *Nature*, 198, 26 (1963).
23. J. D. Roberts and M. C. Caserio, *Basic Principles of Organic Chemistry*, Benjamin, New York, 1964, pp. 86.
24. E. G. Janzen, *Acc. Chem. Res.*, 4, 31, (1971); *Topics in Stereochemistry*, Vol. 6, N. L. Allinger and E. R. Elliel, Eds., Wiley-Interscience, New York, 1971, pp. 177.
25. Ref. 5, pp. 497–500.
26. P. D. Bartlett and T. Funahashi, *J. Am. Chem. Soc.*, 84, 2596 (1962); R. Kuhn and H. Trischmann, *Angew. Chem.*, 75, 294 (1963).
27. M. F. Lappert and P. W. Lednor, *Adv. Organomet. Chem.*, 14, 345 (1976).
28. Ref. 21; G. Sosnovsky and D. Rawlinson, in *Organic Peroxides*, Vol. 2, D. Swern, Ed., Wiley, New York, 1971, pp. 153.
29. H. S. Taylor and H. Diamond, *J. Am. Chem. Soc.*, 57, 1251 (1935).

30. D. H. R. Barton, P. D. Magnus, I. D. Menzies, and R. K. Haynes, *Chem. Commun.*, 511 (1974).
31. A. Nakamura, K. Doi, K. Tatsumi, and S. Otsuka, *J. Mol. Catal.*, 1, 417 (1976).
32. C. A. Tolman, P. Z. Meakin, D. L. Linder, and J. P. Jesson, *J. Am. Chem. Soc.*, 96, 2762 (1974).
33. C. A. Tolman and W. C. Seidel, *J. Am. Chem. Soc.*, 96, 2774 (1974); J. Halpern and T. A. Weil, *Chem. Commun.*, 631 (1973).
34. P. W. Jolly and G. Wilke, *The Organic Chemistry of Nickel*, Vol. 1, Academic Press, New York, 1974, pp. 244–328.
35. L. Malatesta and S. Cenini, *Zerovalent Compounds of Metals*, Academic Press, New York, 1974, pp. 18.
36. M. Matsumoto, H. Yoshioka, K. Nakatsu, T. Yoshida, and S. Otsuka, *J. Am. Chem. Soc.*, 96, 3322 (1974).
37. C. A. Tolman, *J. Am. Chem. Soc.*, 92, 2956 (1970); C. A. Tolmam, W. C. Seidel, and L. W. Gosser, ibid., 96, 53 (1974).
38. T. Yoshida and S. Otsuka, *Inorg. Chim. Acta*, 31, L257 (1978).
39. J. P. Collman and W. R. Roper, *Adv. Organomet. Chem.*, 7, 53 (1968).
40. K. Tatsumi, T. Fueno, A. Nakamura, and S. Otsuka, *Bull. Chem. Soc. Japan*, 49, 2170 (1976).
41. Data taken from C. A. Tolman, *J. Am. Chem. Soc.*, 96, 2780 (1974).
42. J. K. Stille and K. S. Y. Lau, *Acc. Chem. Res.*, 10, 434 (1977).
43. L. Vaska and J. Peone, Jr., *Chem. Commun.*, 1971, 418.
44. D. M. Blake and M. Kubota, *Inorg. Chem.*, 9, 989 (1970).
45. J. Halpern, *Acc. Chem. Res.*, 3, 386 (1970).
46. J. K. Stille, K. S. Y. Lau, and P. K. Wong, *J. Am. Chem. Soc.*, 98, 5832 (1976); ref. 42.
47. J. A. Osborn, in *Organotransition Metal Chemistry*, Y. Ishii and M. Tsutsui, Eds., Plenum Press, New York, 1975, pp. 65.
48. S. Otsuka and K. Ataka, *Bull. Chem. Soc. Japan*, 50, 1118 (1977).
49. A. Nakamura and S. Otsuka, *J. Am. Chem. Soc.*, 95, 7262 (1973).
50. G. Wilke and G. Herrmann, *Angew. Chem.*, 78, 591 (1966); T. Yamamoto, A. Yamamoto, and S. Ikeda, *J. Am. Chem. Soc.*, 93, 3350, 3360 (1971).
51. P. W. Jolly, K. Jonas, C. Krüger, and Y.-H. Tsai, *J. Organomet. Chem.*, 33, 109 (1971).
52. A. Yamamoto, T. Yamamoto, and S. Ikeda, *J. Am. Chem. Soc.*, 93, 3350 (1971).
53. T. Yoshida, T. Yamagata, T. Tulip, J. A. Ibers, and S. Otsuka, *J. Am. Chem. Soc.*, 100, 2063 (1978).
54. L. Malatesta and S. Cenini, *Zerovalent Compounds of Metals*, Academic Press, New York, 1974.
55. H. C. Brown, *Boranes in Organic Chemistry*, Cornell University Press, Ithaca, N.Y., 1972.
56. G. Henrici-Olivé and S. Olivé, *Coordination and Catalysis*, Verlag-Chemie, Weinheim, 1977, pp. 122.
57. J. A. Osborn, F. H. Jardine, J. F. Young, and G. Wilkinson, *J. Chem. Soc., A*, 1966, 1711.
58. T. Miyazawa and T. Ideguchi, *J. Polym. Sci.*, B1, 389 (1963).

Selected Readings

59. J. Cook, W. R. Cullen, M. Green, and F. G. A. Stone, *Chem. Commun.*, 170 (1968).
60. S. Otsuka, M. Naruto, T. Yoshida, and A. Nakamura, *Chem. Commun.*, 396 (1972).
61. W. Hieber and G. Braun, *Z. Naturforsch.*, 12b, 478 (1957).
62. T. Saruyama, T. Yamamoto, and A. Yamamoto, *Bull. Chem. Soc. Japan*, 49, 546 (1976).
63. S. Otsuka, A. Nakamura, and T. Yoshida, *J. Am. Chem. Soc.*, 91, 7196 (1969).
64. F. D. Mango and I. Dvoretzky, *J. Am. Chem. Soc.*, 88, 1654 (1966).
65. (a) K. Noack and F. Calderazzo, *J. Organomet. Chem.*, 10, 101 (1967). (b) R. W. Glyde and R. J. Mawby, *Inorg. Chim. Acta*, 4, 331 (1970); 5, 317 (1971).
66. T. G. Attig and A. Wojcicki, *J. Am. Chem. Soc.*, 96, 262 (1974); *J. Organomet. Chem.*, 82, 397 (1974); T. C. Flood, F. J. Disanti, and D. L. Miles, *Inorg. Chem.*, 15, 1910 (1976).
67. K. S. Y. Lau, P. K. Wong, and J. K. Stille, *J. Am. Chem. Soc.*, 98, 5832 (1976).
68. R. B. Woodward and R. Hoffmann, *The Conservation of Orbital Symmetry*, Academic Press, New York, 1970.
69. R. J. Klinger, W. Butler, and M. D. Curtis, *J. Am. Chem. Soc.*, 97, 3535 (1975).
70. W. I. Bailey, Jr., D. M. Collins, and F. A. Cotton, *J. Organomet. Chem.*, 135, C53 (1977).
71. D. Seyferth, *Adv. Organomet. Chem.*, 14, 97 (1976).
72. M. Ishikawa, T. Fuchikami, T. Sugaya, and M. Kumada, *J. Am. Chem. Soc.*, 97, 5923 (1975); A. G. Brook, J. W. Harris, J. Lennon, and W. M. ElSheikh, *J. Am. Chem. Soc.*, 101, 83 (1979).
73. M. Ishikawa, *Pure Appl. Chem.*, 50, 11 (1978).
74. W. A. Herrmann, *Chem. Ber.*, 111, 1077 (1978); *Angew. Chem. Int. Ed.*, 17, 800 (1978).
75. C. P. Casey and T. J. Burkhardt, *J. Am. Chem. Soc.*, 96, 7808 (1974)
76. R. T. Conlin and P. P. Gasper, *J. Am. Chem. Soc.*, 98, 868 (1976); T. J. Burton, and J. A. Kilgour, *J. Am. Chem. Soc.*, 98, 7231 (1976).
77. R. G. Pearson, *Symmetry Rules for Chemical Reactions*, Wiley-Interscience, New York, 1976.
78. P. Heimbach, in *Aspects of Homogeneous Catalysis*, Vol. 1, R. Ugo, Ed., Manfredi, Milan, 1973.
79. E. L. Muetterties, *MTP International Review of Science, Inorganic Chemistry* Series, Vol. 1, Butterworths, London, 1974; E. L. Muetterties and J. P. Jesson; R. H. Holm, in *Dynamic NMR Spectroscopy*, J. M. Jackman and F. A. Cotton, Eds., Academic Press, New York, 1975, pp. 253 and pp. 317.
80. F. A. Cotton and D. L. Hunter, *J. Am. Chem. Soc.*, 98, 1413 (1976).
81. P. Meakin, E. L. Muetterties, and J. P. Jesson, *J. Am. Chem. Soc.*, 95, 75 (1973).
82. A. Nakamura, to be published.
83. J. W. Reppe, *Ann.*, 601, 81 (1956).
84. R. Jira and W. Freiesleben, *Organomet. React.*, 3, 1 (1972).
85. T. Majima and H. Kurosawa, *Chem. Commun.*, 610 (1977). J. E. Bächvall, B. Åkermark, and S. O. Ljunggren, *J. Am. Chem. Soc.*, 101, 2411 (1979).
86. T. Yoshida, T. Okano, and S. Otsuka, *J. Chem. Soc., Dalton Trans.*, 993 (1976).
87. F. A. Cotton and C. M. Lukehart, *Prog. Inorg. Chem.*, 16, 487 (1972).
88. J. E. Lyons, in *Fundamental Research in Homogeneous Catalysis*, M. Tsutsui and R. Ugo, Eds., Plenum Press, New York, 1977, pp. 1.
89. Z. Dori, D. Getz, E. Melamud, and B. L. Silver, *J. Am. Chem. Soc.*, 97, 3846 (1975).

90. J. Chatt, *J. Organomet. Chem.*, 100, 17 (1975).
91. J. E. Bercaw, in *Fundamental Research in Homogeneous Catalysis,* M. Tsutsui and R. Ugo, Eds., Plenum Press, New York, 1977, pp. 129.
92. S. Otsuka, K. Mori, and K. Yamagami, *J. Org. Chem.*, 31, 4170 (1966).
93. M. Hancock, M. N. Levy, and M. Tsutsui, *Organomet. React.*, 4, 1 (1972); M. Tsutsui and A. Courtney, *Adv. Organomet. Chem.*, 16, 241 (1977).
94. H. H. Zeiss and M. Tsutsui, *J. Am. Chem. Soc.*, 79, 3062 (1957); M. Tsutsui, *Ann. N.Y. Acad. Sci.*, 93, 33 (1961).
95. S. Otsuka, H. Matsui, and S. Murahashi, *Nippon Kagaku Kaishi,* 77, 766 (1956).
96. R. B. Silverman and D. Dolphin, *J. Am. Chem. Soc.*, 98, 4633 (1976); R. H. Abeles and D. Dolphin, *Acc. Chem. Res.*, 9, 114 (1976).
97. K. Vrieze and H. C. Volger, *J. Organomet. Chem.*, 9, 537 (1967).
98. M. Wrigton, G. S. Hammond, and H. B. Gray, *J. Organomet. Chem.*, 70, 283 (1974).
99. M. L. H. Green and P. L. I. Nagy, *J. Organomet. Chem.*, 1, 58 (1963).
100. J. K. P. Ariyaratne and M. L. H. Green, *J. Chem. Soc.*, 2976 (1963).
101. S. Otsuka and A. Nakamura, *Adv. Organomet. Chem.*, 14, 245 (1976).
102. T. Blackmore, M. I. Bruce, F. G. A. Stone, R. E. Davis, and A. Garza, *Chem. Commun.*, 852 (1971).
103. K. Vrieze, in *Dynamic NMR Spectroscopy,* L. M. Jackman and F. A. Cotton, Eds., Academic Press, New York, 1974, pp. 441; J. W. Faller, *Adv. Organomet. Chem.*, 16, 211 (1977).
104. J. Hillis, J. Francis, M. Ori, and M. Tsutsui, *J. Am. Chem. Soc.*, 96, 4800 (1974).

5

MECHANISMS

1. HYDROLYSIS AND CONDENSATION

Hydrolysis and condensation are probably the most important reactions in organic and inorganic chemistry as well as in biological chemistry. A variety of homogeneous catalysts have been utilized, ranging from the simplest one H^+, to metalloenzymes of highly complex structure. For illustrative purposes, we examine acid-catalyzed hydrolysis of organic esters or amides.

Simple protonation of the carbonyl oxygen induces a positive charge on the carbon, which promotes a nucleophilic attack of water on the positive center.

$$\underset{R}{\overset{O}{\underset{\|}{C}}}\text{—}OR' + H^+ \rightleftharpoons \underset{R}{\overset{OH}{\underset{|}{C^+}}}\text{—}OR'$$

Full protonation of an organic acid or ester requires reagents with proton acidity of considerable strength. Approximate pK_a values for such protic acids give a rough measure for the protonation. Thus $HClO_4$ (pK_a −10), HI (pK_a −10), HCl (pK_a −7), and $C_6H_5SO_3H$ (pK_a −7) can fully protonate RCO_2H or

$$\left[R-C\underset{OH}{\overset{OH}{\diagup}}\right]^+ \quad \text{and} \quad \left[R-C\underset{OH}{\overset{OR'}{\diagup}}\right]^+$$

RCO_2R' because the pK_a values of the conjugate acids are larger (~pK −5) than those of most strong acid catalysts. Therefore these strong acids work ef-

109

ficiently for the acid-catalyzed hydrolysis of organic esters. Acid-catalyzed hydrolysis of organic amides, particularly poly- or oligopeptides (i.e., proteins), is an important reaction both biochemically and industrially.[1] The pK_a values of the conjugate acids of organic amides are ≈ -1 and protonation could occur with H_3O^+ (pK_a -1.74). Thus the amides also can be hydrolyzed by a *specific acid-catalyzed process.**

To simplify the acid catalysis discussed above, the presence of nucleophilic substances other than water has been ignored. In actuality under acid hydrolysis conditions, participation of nucleophiles has been found to occur and to accelerate the rate, thus increasing the "efficiency" of the acid catalyst. Usually water has been used as a solvent in these acid hydrolyses, and participation of water or a nucleophile in the transition state of carbonyl protonation by a loose coordination (or nucleophilic interaction) to the carbonyl carbon reduces the energy of the transition state. Therefore a weaker acid (e.g., acetic acid) can partially protonate the carbonyl oxygen with the help of simultaneous nucleophilic interactions. Scheme 12 illustrates an acid-catalyzed hydrolysis.

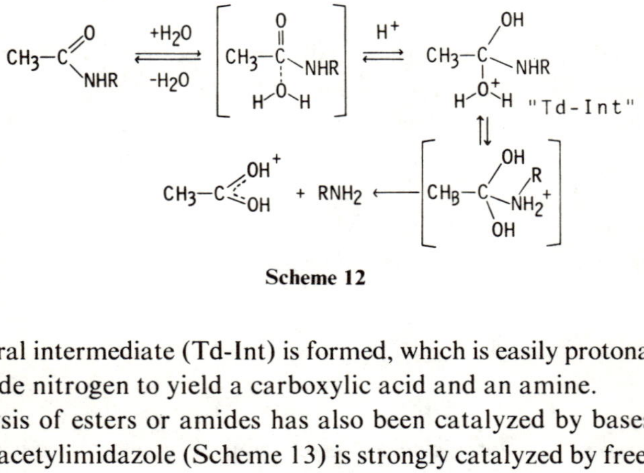

Scheme 12

A tetrahedral intermediate (Td-Int) is formed, which is easily protonated further on the amide nitrogen to yield a carboxylic acid and an amine.

Hydrolysis of esters or amides has also been catalyzed by bases. Base hydrolysis of acetylimidazole (Scheme 13) is strongly catalyzed by free imidazole.

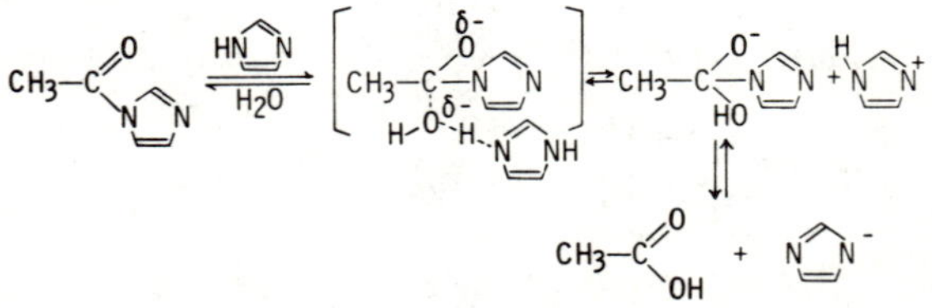

Scheme 13

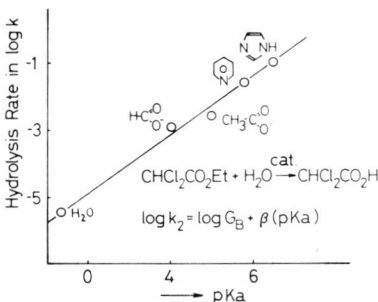

Fig. 47. An example of a linear relationship (Brønsted catalysis law) Between log k_2 and the pK_a of the general base catalysts in the hydrolysis of $CHCl_2CO_2Et$.

The nucleophilic attack by H_2O is assisted by a nucleophilic interaction of imidazole with a proton of the attacking water to give a tetrahedral intermediate as described above. This example illustrates the importance of a general base in assisting base-catalyzed hydrolysis. The plot in Fig. 47 indicates a linear correlation between the rate constant of hydrolysis (k_2) and the strength of the base (pK_a).*

Nucleophilicity toward the carbonyl carbon determines the efficiency of nucleophiles.[2] The following order of nucleophilicity toward the carbonyl carbon has been reported:

$$EtO^- > PhO^- > OH^- > AcO^- > N_3^- > F^- > H_2O > Br^- \sim I^-$$

This order is parallel to the order of pK_a values of the conjugate acids of these nucleophiles. Therefore a carbonyl carbon is a "hard" acid just like H^+.

The similarity in mechanism between both acid- and base-catalyzed hydrolysis is apparent. If protonation and nucleophilic attack occur in concert and in a favorable steric arrangement, the same hydrolysis could be performed in a neutral water solution utilizing specially designed homogeneous catalysts. Some organic esters have been found to hydrolyze even at pH 7 because of suitable steric arrangements (*intramolecular catalysis*) as shown in Scheme 14.[3]

The importance of the steric arrangement of nucleophile relative to the carbonyl group has been demonstrated[4] in a kinetic study of the lactonization (internal esterification) of neatly designed hydroxyacids (Table 13).

* $pK_a = \dfrac{[B][H^+]}{[BH^+]}$ in equilibrium $BH^+ \rightleftharpoons B + H^+$

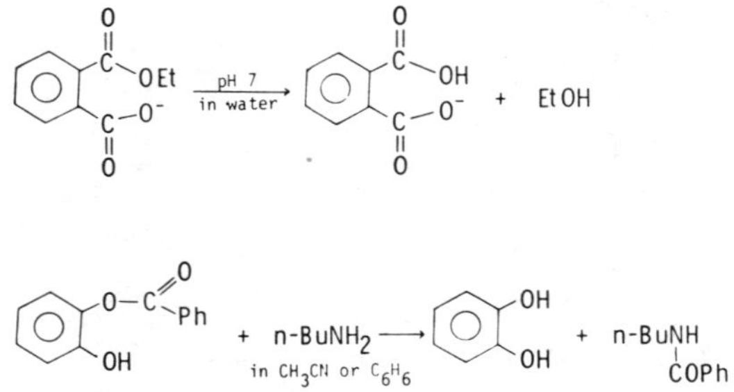

Scheme 14

The most favorable geometry for concerted electrophilic and nucleophilic interactions on normal esters or amides is as follows:

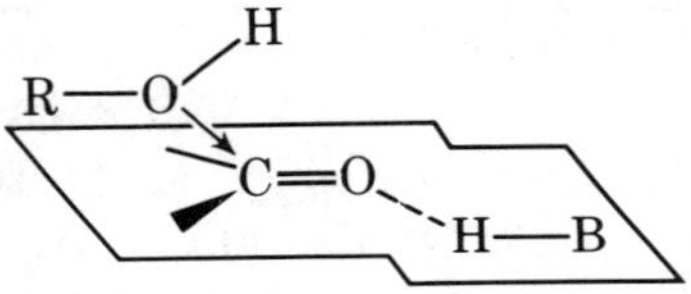

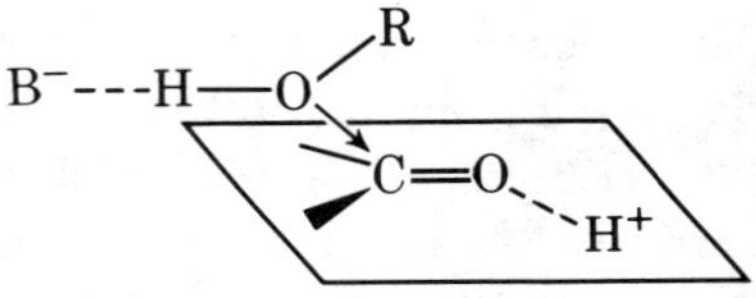

Small organic molecules are unable to effect these concerted interactions. A sufficiently large molecule with suitable crevices or holes that just fit the esters or amides has been generally required. Hydrolytic enzymes meet this requirement and work ideally at neutral conditions (pH ∼ 7) at higher rates than those observed with very strong acids or bases as catalysts. Proteins are rapidly hydrolyzed by hydrolytic enzymes (called proteases) at pH ∼ 7. The high efficiency of these enzymes has been explained by suggesting that the cooperative action

Table 13. Relative Rates of Lactonization (Rate of Ethyl Acetate Is Taken as Unity)

compound	rel. rate	compound	rel. rate
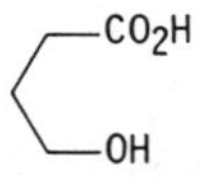 -CO₂H / -OH	79.5	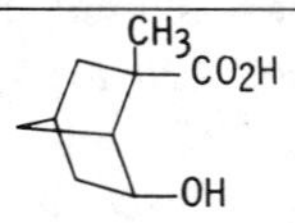 CH₃, -CO₂H / -OH	276
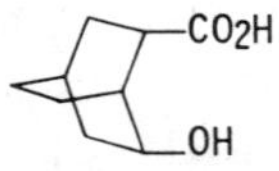 -CO₂H / -OH	871	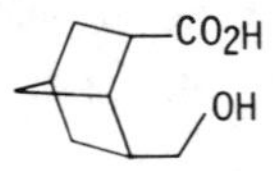 -CO₂H / -OH	6,621
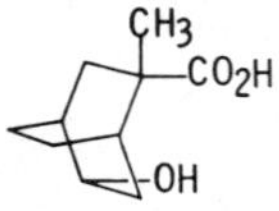 CH₃, -CO₂H / -OH	479,000	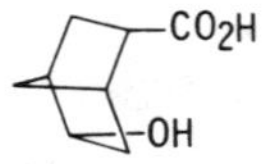 -CO₂H / -OH	1,030,000

Hydrolysis and Condensation

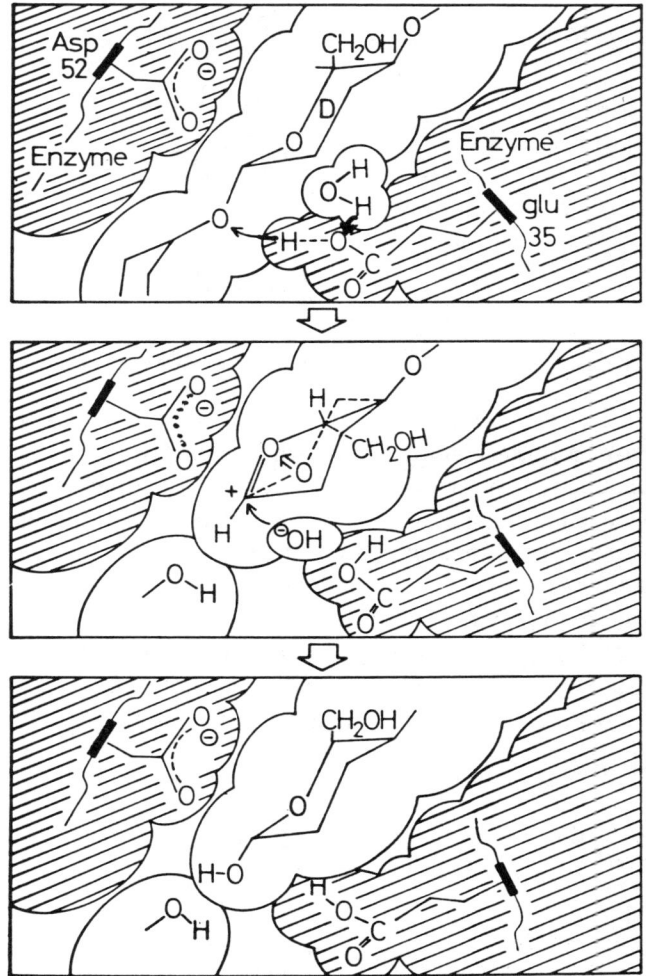

Fig. 48. An illustration of lysozyme action.

of locally acidic as well as basic sites just fits the amide bond being hydrolyzed. As an example of enzyme action, the hydrolysis of an acetal linkage by lysozyme is illustrated in Fig. 48.[5] The special substrate (a tetraglycolide) just fits in the crevice of lysozyme by forming several sterically oriented hydrogen bonds. As water penetrates the enzyme channel, it interacts with the —CO_2H group of glutamic acid (Glu 35). The water molecule facilitates protonation of the acetal oxygen by the proton of Glu 35. During this process the water is ionized to a

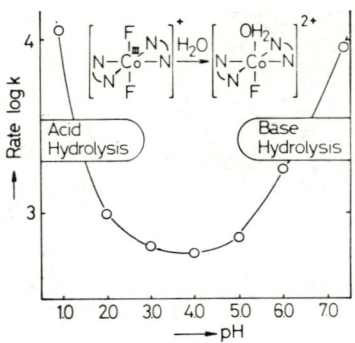

Fig. 49. The dependence on pH of the hydrolysis rate of *trans*-[CoF$_2$(en)$_2$]$^+$ at 59.3°.

hydroxide ion by the resulting —CO$_2^-$ group of Glu 35. At the same time, a positive charge is formed on the acetal carbon, which is then stabilized by the —CO$_2^-$ group of aspartic acid (Asp 52). The hydroxyl attacks the carbonium ion in a nucleophilic manner and completes the hydrolysis. The enzyme then relaxes and liberates the hydrolyzed substrates.

Substitution reactions of metal complexes have been also catalyzed by homogeneous acids and bases. For example, hydrolytic substitution of the F$^-$ ligand in *trans*-[CoF$_2$(en)$_2$]$^+$ by water (aquo ligand) is accelerated by H$^+$ and OH$^-$. The observed pH dependence of the rate is shown in Fig. 49.[6]

The proposed mechanism for the acid-catalyzed reaction is as follows:

Overall reaction:

$$[F-CoF(en)_2]^+ + H_2O \xrightarrow[k_1]{\text{very slow}} [H_2O-CoF(en)_2]^{2+} + F^-$$

Mechanism:

$$[F-CoF(en)_2]^+ + H^+ \underset{K_{aq}}{\rightleftharpoons} [HF-CoF(en)_2]^{2+}$$

$$\xrightarrow{k_2} [H_2O-CoF(en)_2]^{2+} + HF \quad (+ H_2O)$$

Rate equation:

$$\text{rate} = k_1[\text{CoF}_2(\text{en})_2^+] + k_2 K_{aq}[\text{CoF}_2(\text{en})_2^+][H^+]$$

Electrophiles such as Hg^{2+}, Tl^{3+}, or Ag$^+$ also accelerate similar halide substitution reactions by a S_E2 (electrophilic bimolecular substitution) mechanism. With metal complexes in particular, bimolecular reactions or mechanisms have usually been called "associative reactions" or "associative mechanisms."

Polar Addition and Ionic Elimination

Metal ions promote many polar reactions: electrophilic catalysis, superacid catalysis, and metal ion catalysis. Hydrolysis of amino acid esters has been efficiently catalyzed with Mg^{2+}, Ca^{2+}, Mn^{2+}, Co^{2+}, and Cu^{2+} at pH 7 to 8. The trend in activity of metal ions parallels the formation constants of ammine complexes ($Cu^{2+} > Co^{2+} > Mn^{2+} > Ca^{2+}$). The mechanism of Scheme 15 has been supported by an ^{18}O isotopic experiment.[7]

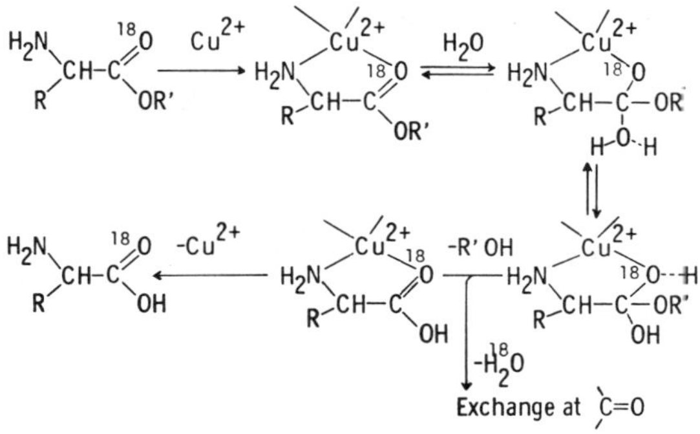

Scheme 15

Metal ions also work in the hydrolysis of condensed phosphates. A relevant biochemical example is the hydrolysis of ATP catalyzed by Ca^{2+} or Cu^{2+}:

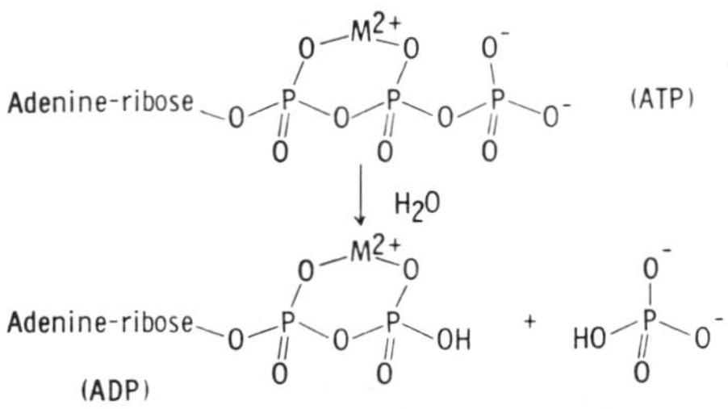

Peptide linkages are hydrolytically cleaved by catalysis with M^{2+}. Certain metalloenzymes (e.g., leucine amino peptidase) accelerate a similar cleavage of a particular peptide bond by action at an electrophilic metal ion site.

2. POLAR ADDITION AND IONIC ELIMINATION

Addition and elimination are the most important elementary steps in chemical reactions. A general description of these two reaction types is too vast a topic to be covered here, and only the reactions that occur through essentially ionic mechanisms under a *polar* influence of reagents or catalysts are described.

Polar additions of water to unpolarized unsaturation such as found in R—C≡C—R or RCH=CHR have been catalyzed by electrophilic metal ions or Lewis acids. Mercuric salts have been well known for catalyzing the hydration of acetylenes and the addition of hydrogen chloride to acetylenes. The intermediacy of the mercurinium ion

$$\begin{array}{c} HC \\ \parallel \\ HC \end{array}\!\!\!\!>\!\!Hg^{2+}$$

is probable. This ion then undergoes a σ-π rearrangement (Chapter 4, Section 2.7) to give a cationic σ-alkenylmercuric complex, which is decomposed by protolytic cleavage of the Hg—C bond.

$$HC \equiv CH + H_2O \xrightarrow{Hg^{2+}} \left[\begin{array}{c} H-C \\ \oplus \\ H-C \end{array}\!\!\!Hg \cdot L \right]^{2+} \rightleftarrows \left[\begin{array}{c} H \\ C \\ \vert \\ C \\ H \end{array}\!\!\!Hg \cdot L \right]^{2+} \rightarrow CH_3CHO$$

$$HC \equiv CH + HCl \xrightarrow{HgCl_2} CH_2 = CHCl$$

Hydration of an olefinic bond has been homogeneously catalyzed by strong protic acids (e.g., concentrated H_2SO_4, HF, HCO_2H), but the use of "solid acids" in a heterogeneous phase has been the preferred method in industry. Obviously the initial protonation of the double bond to give a carbonium ion is the important catalytic step.

$$\begin{array}{c} R \\ \diagdown \\ H \end{array}\!\!\!C = CH_2 + H_2O \xrightarrow{H^+} \begin{array}{c} R \\ \diagdown \\ CH-CH_3 \\ \vert \\ OH \end{array}$$

The addition of alcohols to a ketonic group is an acid-catalyzed reaction for which the following path has been suggested:

$$>\!\!C=O \underset{}{\overset{H^+}{\rightleftarrows}} >\!\!\overset{+}{C}\!\!-\!\!OH \underset{}{\overset{ROH}{\rightleftarrows}} \left[>\!\!C\!\!<\!\!\begin{array}{c} H \\ OR \\ OH \end{array} \right]^+ \xrightarrow{-H_2O}$$

$$>\!\!\overset{+}{C}\!\!-\!\!OR \underset{}{\overset{ROH}{\rightleftarrows}} \left[>\!\!C\!\!<\!\!\begin{array}{c} -OR \\ \vert \\ HOR \end{array} \right]^+ \xrightarrow{-H^+} >\!\!C\!\!<\!\!\begin{array}{c} OR \\ OR \end{array}$$

Polar Addition and Ionic Elimination

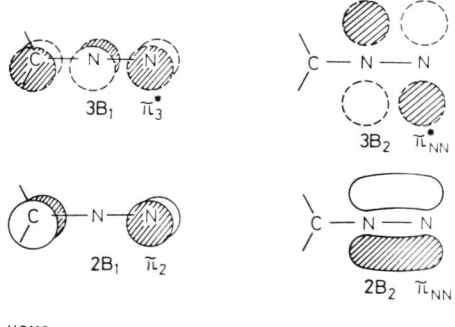

HOMO

Fig. 50. The shapes and orbital symmetries of important molecular orbitals of diazomethane.

The mechanism is similar to that of the acid-catalyzed hydrolysis of an ester as described above.

Lewis acids ($ZnCl_2$, $LiCl$, $TiCl_4$, etc.) work as catalysts for the catalytic decomposition of diazo compounds.[8] The interaction of the partially anionic carbon

$$2RCHN_2 \xrightarrow{\text{Lewis acid}} RCH = CHR + 2N_2$$

of the diazo compound with the Lewis acid center has been thought to weaken the C—N bond and induce dinitrogen evolution. In considering the molecular orbitals in this process (Fig. 50), we see that the positive overlap of the orbital with a relevant vacant metal orbital will reduce the bonding contribution in the C—N bond. Thus there has been an attempt to explain the influence of a Lewis acid on the catalytic decomposition of diazo compounds by MO theory.[9]

The interaction of $2B_1$ (HOMO) with the vacant metal orbital weakens the C—N bond, thereby releasing an N_2 molecule.

The decarboxylation of β-ketoacids is an example of protons and metal ions, as well as organic amines or coenzymes, working as catalysts. The electrophilic character of suitable metal ions seems to stabilize the intermediate stage of the decarboxylation by serving as an "electron sink." The observed trend in catalytic activity supports the view above following the order of activity: $Al^{3+} > Fe^{3+} > Cu^{2+} > Fe^{2+} > Zn^{2+} > Mg^{2+}$. A mechanism involving organic amines is shown below:

Here also, the partial positive charge on the nitrogen in the Schiff base type of intermediate helps to stabilize the developing negative charge on the —CH_2 during CO_2 elimination from —CH_2—CO_2^- group.

Carbonic anhydrase is a well-known enzyme for the elimination of CO_2 from HCO_3^- (Fig. 51). The electrophilicity at the zinc center in a suitable steric orientation seems to be an important factor.[10] Interestingly, the activity trend in changing the identity of metal ions in the carbonic anhydrase (i.e., $Zn^{2+} \gg Co^{2+}$

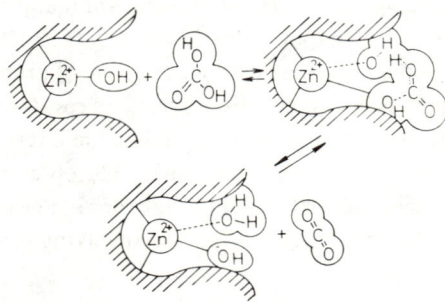

Fig. 51. A schematic representation of the mechanism of carbonic anhydrase action.

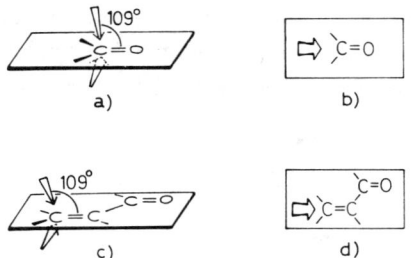

Fig. 52. Preferred orientations of nucleophilic attack on (*a, b*) a carbonyl and (*c, d*) an activated double bond.

> Ni^{2+} > Cu^{2+}) does not follow the general trend for coordination of nucleophiles to first-row transition metal ions (Cu^{2+} > Ni^{2+} > Co^{2+} > Zn^{2+}) usually called the Irving-Williams series.

The simple organic polar addition-elimination reaction is especially accelerated when polar attack of an electrophile or nucleophile occurs in a unique direction favored by the substrate.[11] For example, in polar nucleophilic addition to a ketonic carbonyl group, the most favored directions are as shown in Fig. 52.

Precise control of the direction of attack is impossible for the usual reagents. However the unique feature of an enzyme in accelerating these polar reactions is the precise control of direction achievable by fixing the substrate in the right position relative to the attacking species.

3. ELECTRON TRANSFER REDOX REACTIONS

As described earlier, electron transfer is a key step in many redox reactions. However this phenomenon does not always occur rapidly enough. Catalysts that speed up the electron transfer step have been usually utilized. For example, chemical methods of reductive cleavage of dinitrogen (N_2) require multielectron transfer from strong electron donors (e.g., sodium metal). Direct reaction between dinitrogen and sodium metal does not proceed very fast at ambient conditions. A combination of naphthalene and tetraalkoxytitanium(IV) in THF is a well-known catalyst for this electron transfer process[12]:

Biological nitrogen fixation has been found to utilize much more sophisticated systems of electron transfer. Details remain unknown, but a probable electron transfer pathway is represented in Scheme 16, where FAD is flavin adenine dinucleotide.

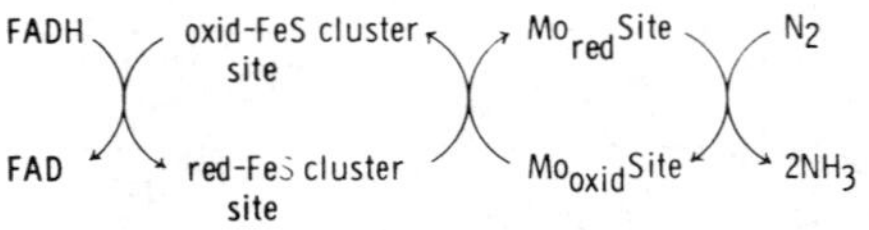

$$\text{FADH} \diagdown \text{oxid-FeS cluster site} \diagdown \text{Mo}_{red}\text{Site} \diagdown \text{N}_2$$
$$\text{FAD} \diagup \text{red-FeS cluster site} \diagup \text{Mo}_{oxid}\text{Site} \diagup 2\text{NH}_3$$

Scheme 16

It is important to note that a proton transfer pathway is closely coupled with electron transfer pathways in these biochemical mechanisms.

The electron transfer chain (Scheme 17) is of vital importance in biochemical mechanisms. In particular, metabolism depends on such an electron-transporting device sometimes called a respiratory chain.

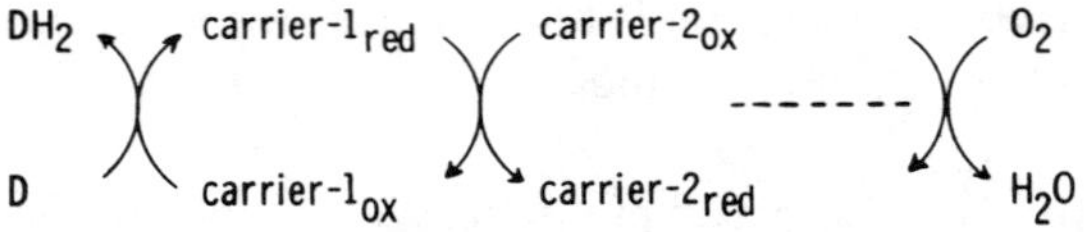

$$\text{DH}_2 \diagdown \text{carrier-1}_{red} \diagdown \text{carrier-2}_{ox} \diagdown \cdots \diagdown \text{O}_2$$
$$\text{D} \diagup \text{carrier-1}_{ox} \diagup \text{carrier-2}_{red} \diagup \cdots \diagup \text{H}_2\text{O}$$

Scheme 17

Examples of the carrier are cytochromes (a, b, c, and c_1), ferredoxins, nicotine-adenine-dinucleotide (NAD), FAD, and molybdenum active sites of relevant Mo enzymes.[14]

Electron transfers among these specific carriers are specifically accelerated by certain redox enzymes (oxidoreductases), for example, cytochrome c oxidoreductase or NADH reductase. Details of these biochemical electron transfers still have not been elucidated. Recent progress in the fields of metalloporphyrins, metal clusters, micelle systems, and macromolecular systems should provide a better understanding of these vitally important processes.

The ability of one chemical substance to oxidize or reduce another by electron transfer has been quantitatively expressed by the redox potential E_0 under specified conditions.

$$\text{observed potential} = E = E_0 - \frac{RT}{nF} \ln Q$$

where n = number of electron transferred
F = Faraday's constant
Q = concentration quotient of participating redox pairs

Electron Transfer Redox Reactions

Comparing the E_0 values of the ions or systems participating in electron transfer yields the direction of the transfer and its electrical feasibility. For example, the aquated Fe^{3+}/Fe^{2+} system has an E_0 value of $+0.77$ V, whereas that of the $O_2 + 2H^+/H_2O$ system (at pH 7) is $+0.82$ V, this indicates that complete oxidation of Fe^{2+} to Fe^{3+} by O_2 at pH 7 will be difficult. However, when Fe^{2+} is surrounded by special ligands, as in cytochrome a (an electron carrier), the E_0 value is $+0.29$ V. With the aid of cytochrome oxidase, the electron transfer reaction with O_2 smoothly occurs to give the oxidized cytochrome a (Fe^{3+}). Delicate and complex networks of electron transport systems in biological redox reactions utilize O_2 as the final electron acceptor. As yet such complex electron transfer systems are not widely utilized in the chemical oxidation of biochemically important materials in which O_2 serves as an inexpensive oxidant.

Simple transition metal ions sometimes strongly catalyze redox reactions. Copper(I, II), silver(I, II), iron(II, III), molybdenum(IV, V, VI) and manganese(II, III) ions are among these catalytic ions. A homogeneous solution of Cu(I) catalyzes the oxidation of CN^- by $[Fe^{III}(CN)_6]^{3-}$, and the oxidation of V(III) ion by Fe(III):

$$CN^- + [Fe^{III}(CN)_6]^{3-} \xrightarrow{Cu^+} CNO^- + [Fe^{II}(CN)_6]^{4-}$$

$$[V^{III}(H_2O)_6]^{3+} + [Fe^{III}(H_2O)_6]^{3+} \xrightarrow{Cu^+} [V^{IV}(H_2O)_6]^{4+} + [Fe^{II}(CN)_6]^{2+}$$

Direct electron transfer between substrate and the metal ion (e.g., V^{3+} and Fe^{3+}) is very slow—that is, requires relatively high activation energy—mainly because of the high formal charges on both reactants. In contrast, electron transfer between these metal ions in the presence of Cu^+ or Cu^{2+} readily occurs, as follows:

$$V^{3+} + Cu^{2+} \xrightarrow{\text{one-electron transfer}} V^{4+} + Cu^+$$

$$Fe^{3+} + Cu^+ \xrightarrow{\text{one-electron transfer}} Fe^{2+} + Cu^{2+}$$

Exchange between the two oxidation states of Tl^+ and Tl^{3+} ($Tl^{*3+} + Tl^+ \rightarrow Tl^{3+} + Tl^{*+}$), as studied by isotopic labeling, is catalyzed by the Fe^{2+} ion, which works as a one-electron transfer agent thus[15]:

$$Tl^{3+} + Fe^{2+} \rightleftharpoons Tl^{2+} + Fe^{3+}$$

$$Tl^{2+} + Fe^{2+} \rightleftharpoons Tl^+ + Fe^{3+}$$

The action of copper ions is also highly modified by cooperation with relevant enzymes. Various oxidases or oxygenases contain copper ions in their active centers where unique selective oxidation is smoothly catalyzed. For example, tyrosinase (a Cu-containing oxidase) catalyzes the air oxidation of phenylalanine

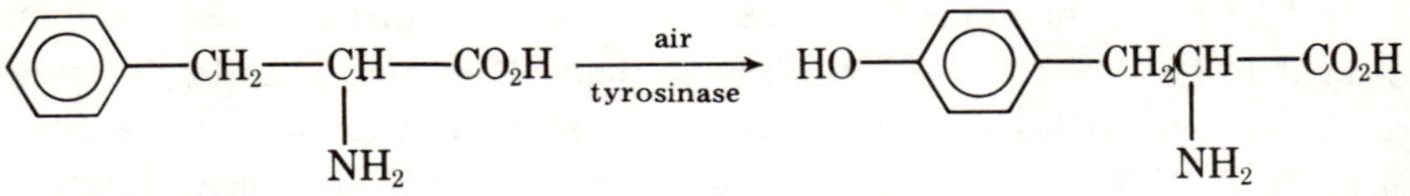

$$\text{C}_6\text{H}_5\text{—CH}_2\text{—CH(NH}_2\text{)—CO}_2\text{H} \xrightarrow[\text{tyrosinase}]{\text{air}} \text{HO—C}_6\text{H}_4\text{—CH}_2\text{CH(NH}_2\text{)—CO}_2\text{H}$$

to tyrosine. Electron transfers between these reactants are thus effectively mediated by a Cu ion in a suitable environment. Similarly, hydroxylation of hydrocarbons is catalyzed by iron-heme enzymes, in particular P-450, where the cooperative electron transfer action of the Fe center, the surrounding porphyrin ligand, and the axial organic sulfur ligand seems to be important.

4. GROUP TRANSFER REACTIONS

Amino group transfer (transamination) has been efficiently catalyzed by certain metal ions or transaminases. Scheme 18 shows a mechanism involving chelate formation with a dipositive metal (L = ligand). This mechanism also explains the rapid racemization of an α-amino acid in the presence of a dipositive metal ion and salicylaldehyde, which functions the role of R—CO—CO$_2^-$. Amino group transfer can be catalyzed by another pyridoxal phosphate that is a typical coenzyme.

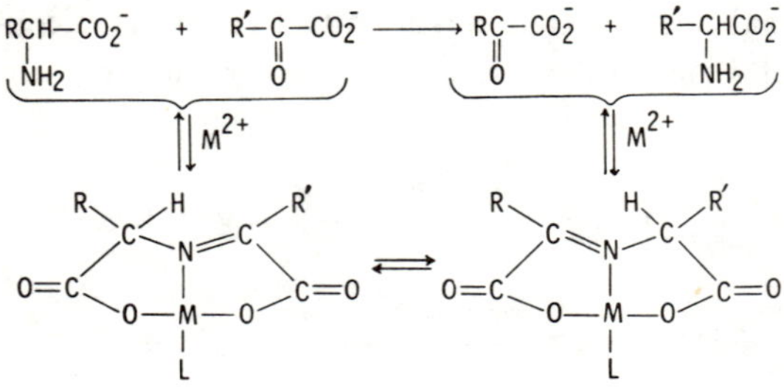

$$\text{RCH(NH}_2\text{)—CO}_2^- + \text{R'—C(=O)—CO}_2^- \longrightarrow \text{RC(=O)—CO}_2^- + \text{R'—CH(NH}_2\text{)CO}_2^-$$

$$\Updownarrow M^{2+} \qquad\qquad\qquad \Updownarrow M^{2+}$$

[chelate intermediates with M—L]

Scheme 18

5. METHYL TRANSFER REACTIONS

The methyl group on a betaine is transferred to L-homocysteine to give L-methionine by S-methyltransferase.

$$(CH_3)_3N^+\text{—}CH_2CO_2^- + HSCH_2CH_2\underset{\underset{NH_2}{|}}{CH}\text{—}CO_2H$$

$$\downarrow$$

$$(CH_3)_2NCH_2CO_2H + CH_3SCH_2CH_2\underset{\underset{NH_2}{|}}{CH}\text{—}CO_2H$$

Apparently the methyl group on the quaternary nitrogen is transferred to some nucleophilic part of the enzyme and the S-methylation then proceeds.

Methylcobalamine or methylcorrinoid is involved in various biologically important methyl transfer reactions because of the exceptionally high nucleophilicity (supernucleophilicity) of natural cobalamin(I) [synthetic cobaloxime(I) is also highly nucleophilic]. The intermediacy of the organometallic methylcobalt species is very important for methyl group transfer reactions of methylcobalamin. Methyl transfer to carbon dioxide to give acetic acid is one of the most important reactions in biosynthesis. N-Methyltetrahydrofolic acid and methylcobalamin cooperate in this process:

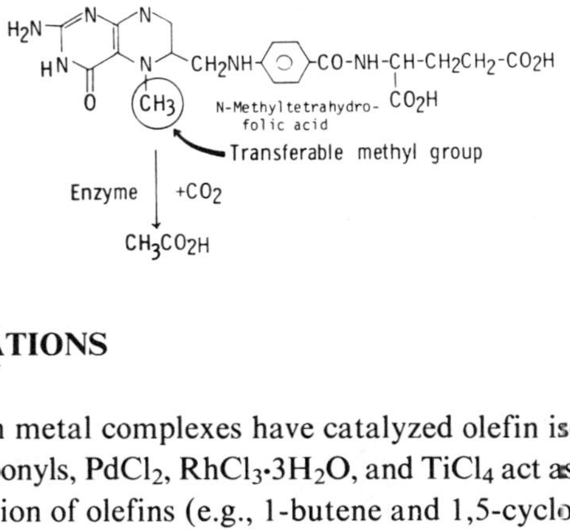

6. ISOMERIZATIONS

Various transition metal complexes have catalyzed olefin isomerizations. For example, iron carbonyls, $PdCl_2$, $RhCl_3 \cdot 3H_2O$, and $TiCl_4$ act as efficient catalysts for the isomerization of olefins (e.g., 1-butene and 1,5-cyclooctadiene):

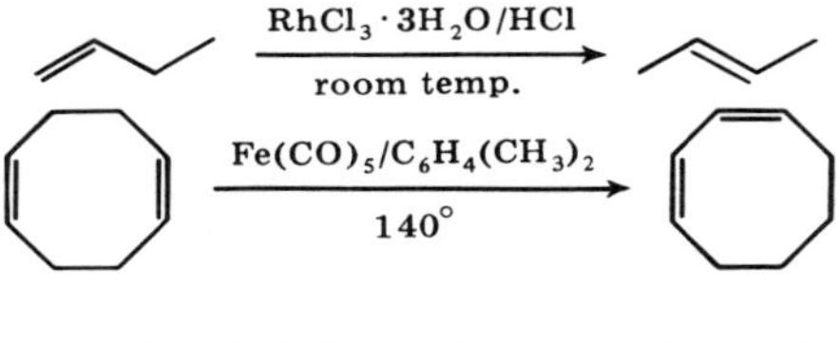

The proposed mechanism for olefin isomerization catalyzed by rhodium complex involves the insertion of an olefin into a Rh—H bond to form a labile alkylrhodium species, which then reverts to the rhodium hydride, releasing the isomerized olefin by β elimination[16]:

$$[Rh^I]^\circ + HCl \longrightarrow [RhHCl]^\circ$$

$$[RhHCl]^\circ + \text{/\!\!\!\!\!\diagup\!\!\!\diagdown} \rightleftharpoons [RhCl]^\circ$$

$\circ$: The auxiliary ligands are omitted for clarity

$$[HRhCl]^\circ + \text{/\!\!\!\!\!\diagup\!\!\!\diagdown}$$

In the case of an iron carbonyl catalyst, the isomerization proceeds by the formation of a labile iron carbonyl hydride active catalyst, which is formed by abstraction of hydrogen from the olefin.

In the case of Pt(II), a proposed intermediate in the insertion/de-insertion step has been isolated as a hydrido-olefin complex $[PtH(C_2H_4)(PEt_3)_2]^+$ by the low-temperature reactions of $[PtH(acetone)(PEt_3)_2]^+$ with C_2H_4.

Even when there is apparently no double bond isomerization hydrogen isotopic exchange occurs via a rapid insertion/de-insertion process[17]:

$$R-CH=CH_2 + MD(L)_n \longrightarrow [R-\underset{D}{CH}-CH_2-M(L)_n]$$

$$R-CD=CH_2 + MH(L)_n \longleftarrow$$

Catalytic isomerization of internal olefins $RCH=CHR'$ to terminal olefins (e.g., $R'-CH=CH_2$) has not been achieved because the internal olefins have greater thermodynamic stability than the terminal ones. However with a stoichiometric quantity of certain boron or aluminum hydrides, insertion products (branched alkyls) from the internal olefins can be converted to the corresponding normal alkyl metals by heating. Thermal β elimination of the n-alkyl groups yield terminal olefins:

$$\underset{CH_3}{\overset{R}{\diagdown}}CH=CH + \diagdown BH \xrightarrow{\text{insertion}} \underset{-B\ \ CH_3}{\overset{R-CH_2}{\diagdown}CH}$$

$\downarrow$ metallotropic rearrangement by heating

$$\underset{CH=CH_2}{\overset{RCH_2}{\diagdown}} + \diagdown BH \xleftarrow{\substack{\beta\text{-elimination} \\ \text{(de-insertion)}}} \underset{}{\overset{RCH_2}{\diagdown}}CH_2-CH_2-B\diagup$$

Hydrogenations

Skeletal isomerization of some 1,4-dienes has been found to occur with some catalyst systems [e.g., $NiCl_2(PEt_3)_2/AlEt_3$] at room temperature. The mechanism of Scheme 19 is for the formation of skeletally isomerized 1,3-dienes-d_2 from 2-methyl-1,4-pentadiene-1-d_2.[18,19]

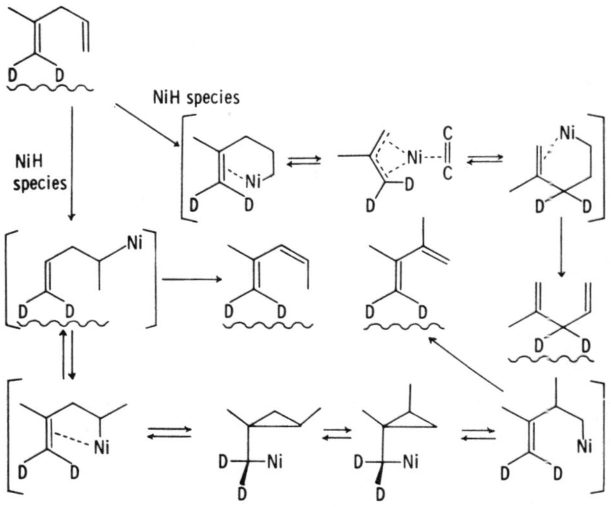

Scheme 19

It has been found that skeletal rearrangement of strained small ring organic compounds can be catalyzed by an extensive variety of transition metal ions (Ag^+) or neutral complexes [$PdCl_2(PhCN)_2$, $Rh_2Cl_2(CO)_4$, or Ni-$(CH_2=CHCN)_2$][20]:

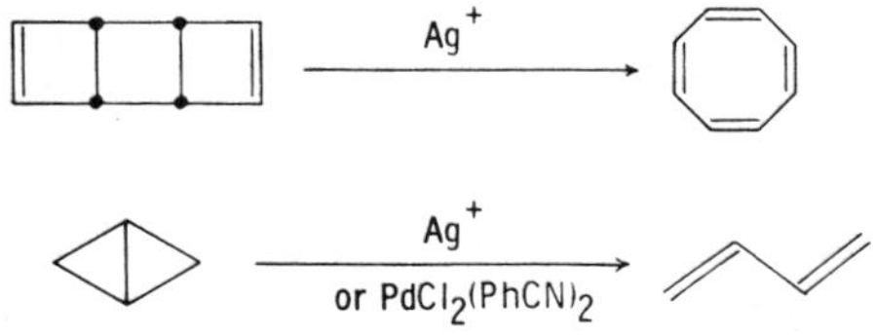

Mechanisms involving orbital symmetry relaxation have been proposed.

7. HYDROGENATIONS

Homogeneously catalyzed hydrogenation of olefins with aqueous solutions of Ag^+, Cu^+, Ru^{2+}, or $Co(CN)_5^{3-}$ has long been known but has not been widely used because of the poor catalytic activity of these metal ions. In 1966 Wilkinson

Table 14. Typical Examples of Homogeneous Hydrogenation Catalysts[21]

Metal Complexes	Mechanism	Characteristics
$RhCl(PPh_3)_3$, the "Wilkinson complex"	η^4	Highly active for mono- or disubstituted olefins and acetylenes
$RhH(CO)(PPh_3)_3$	η^3	Selectively active for monosubstituted olefins; inert to acetylenes
$RuCl_2(PPh_3)_3$	η^3	Active in C_6H_6/EtOH mixture; active species is $RuHCl(PPh_3)_3$
$RuHCl(PPh_3)_3$	η^3	Very active for terminal olefins
$[Co(CN)_5]^{3-}$	η^3	Weakly active; monohyrogenation of acetylenes or conjugated dienes

$[RhCl(C_2H_4)_2]_2/P(CH_3)(C_6H_{11})Ph$		Highly enantioselective catalyst system for PhCH=C—CO$_2$H
		                    |
		                   NHCOCH$_3$

and his group reported the very high catalytic activity of $RhCl(PPh_3)_3$ toward hydrogenation of various types of olefin in organic solvents. The activity exceeds that of the well-known platinum oxide in the hydrogenation of cyclohexene at 20° under 1 atm of H_2. Since this discovery a number of similar noble metal complexes have been examined, and some compounds have been found to have high activity as well as excellent regio- or enantioselectivity. Table 14 gives typical examples of these hydrogenation catalysts.

The mechanism involving an initial oxidative addition of H_2 to a low-valent metal is shown in Scheme 20. Since the most important catalytic intermediate involves a η^2-olefin and two η^1-H ligands, this hydrogenation has been called tetrahapto(η^4)-hydrogenation. Another hydrogenation mechanism involves olefin insertion into an M—H bond followed by hydrogenolysis of the resulting metal alkyl as shown below. This mechanism has been called trihapto(η^3)-hydrogenation:

$$HML_n \xrightarrow{\!\!>\!\!C=C\!\!<\!\!} HM\!\!\left(\!\!\begin{array}{c}\!\!>\!\!C=C\!\!<\!\!\end{array}\!\!\right)\!\!L_n$$

(η^2-olefin-η^1-hydrido-intermediate)

$$>\!\!CH\!-\!CH\!\!<$$

$$>\!\!CH\!-\!\underset{|}{C}\!-\!M(H_2)L_m \xleftarrow{H_2} >\!\!CH\!-\!\underset{|}{C}\!-\!ML_n$$

(di-η^1-Hydrido-η^1-alkyl complex)   (η^1-Alkyl complex)

The stereochemistry of these homogeneous hydrogenations has generally been found to be cis, provided geometrical isomerization of the reactants or the products does not occur, obscuring the overall stereochemistry.

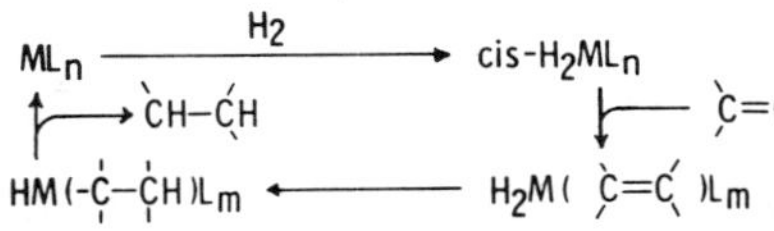
H2
MLn
cis-H2MLn
CH–CH
C=C
HM(-C–CH)Lm
H2M(C=C)Lm

Scheme 20

The reaction paths and the stereochemistry of rhodium phosphine "Wilkinson" complexes are governed by the trans effect of ligands involved in the catalysis.[21b] The oxidative addition of H_2 occurs cis to give dihydrido tris(phosphine) complex, which has one unique phosphine ligand trans to one of the hydrido ligands. This particular phosphine readily dissociates in solution and is displaced with olefin, giving dihydrido(olefin) species. This stage is very unstable and immediately is transformed into a *cis*-hydridoalkyl complex, which eliminates alkane to complete the olefin hydrogenation. Two phosphine ligands are always retained at mutually a trans position throughout this catalytic cycle. The high trans effect of hydrido- and alkyl ligands destabilizes *trans*-dihydrido or *trans*-hydridoalkyl complexes, which is why the above-mentioned reaction paths are favored.

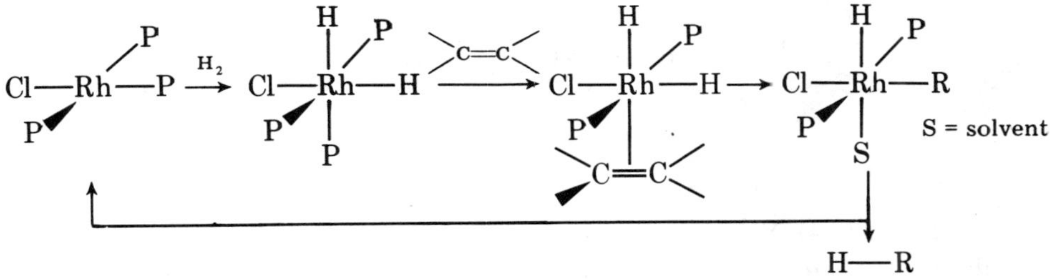
Cl—Rh—P
H2
Cl—Rh—H
C=C
Cl—Rh—H
Cl—Rh—R
S = solvent
H—R

Ketones or aromatic rings have also been hydrogenated by some homogeneous catalysts. Proper selection of metal species and ligands is crucial for these hydrogenations.[22,23]

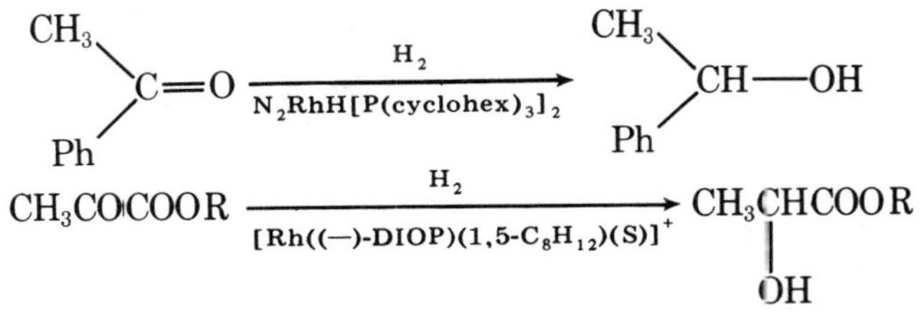
CH3
C=O
Ph
H2
N2RhH[P(cyclohex)3]2
CH–OH
CH3COCOOR
[Rh((—)-DIOP)(1,5-C8H12)(S)]+
CH3CHCOOR
OH

$$\underset{\underset{O}{\parallel}}{\underset{CH_3}{\overset{CH_3}{\diagdown}}}\!\!\!\!\!\!\!\!\!\!\!\!\!\!\diagup\!\!\!\!\diagdown_{O} \xrightarrow[S=\text{solvent}]{H_2 \atop [Rh((-)\text{-DIOP})(1,5\text{-}C_8H_{12})(S)]^+} \underset{\underset{O}{\parallel}}{\underset{HO}{\overset{CH_3}{\diagdown}}}\!\!\!\!\!\!\!\!\!\!\!\!\!\!\diagup\!\!\!\!\diagdown_{O}$$

Pantolactone

$$\text{Ph-R} \xrightarrow[{[Rh(Me_5Cp)X_2]_2}]{H_2 \atop Co(\eta^3\text{-}C_3H_5)[P(OPh)_3]_3} \text{Cy-R}$$

Hydrolysilylation of olefins or ketones also proceeds homogeneously with RhCl(PPh$_3$)$_3$ as the catalyst. Oxidative addition of Si—H bonds occurs prior to olefin insertion into the resulting M—H bond.

The hydrosilylation of ketones gives alkoxysilanes that form the corresponding alcohols on hydrolysis with methanol. This overall reduction of ketones has been called a "hydrosilylation reduction." By utilizing relevant optically active phosphines, asymmetric synthesis of some secondary alcohols has been possible

$$R_3SiH + \underset{}{\overset{}{>}}C=C\underset{}{\overset{}{<}} \xrightarrow{RhCl(PPh_3)_3} R_3Si-\underset{|}{\overset{|}{C}}-\overset{|}{C}H$$

$$\Bigg\downarrow RhCl(PPh_3)_3 \Big| \underset{}{\overset{}{>}}C=O$$

$$R_3Si-O-CH\underset{}{\overset{}{<}} \xrightarrow{MeOH} R_3SiOCH_3 + H-O-CH\underset{}{\overset{}{<}}$$

$$\text{C}_6\text{H}_6 + H_2 \xrightarrow{HD/SbF_5} \text{C}_6\text{H}_7\text{-H}$$

(see Section 14). Very strongly acidic systems (e.g., HF/TaF$_5$ or HF/SbF$_5$) also hydrogenate aromatic compounds at 50° under 50 to 100 atm of H$_2$[24]:

The mechanism probably involves a series of processes in which protonation is followed by addition of hydride:

$$\text{C}_6\text{H}_6 + H^+ \longrightarrow [\text{C}_6\text{H}_7]^+ \xrightarrow{H^-} \text{C}_6\text{H}_8$$

$$H_2 \longrightarrow H^- + H^+$$

Hydrogenations

The hydride results from heterolytic cleavage of molecular hydrogen in the extremely polar reaction medium. In support of the mechanism, hydride can also be generated from the heterolytic cleavage of C—H bonds at tertiary carbons under the same conditions[24]:

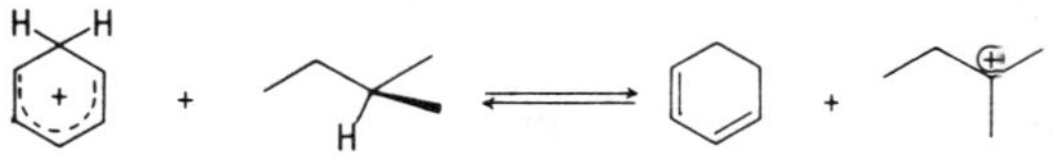

Some polyaromatic compounds are hydrogenated by $Co_2(CO)_8$ at 150° under a H_2/CO mixture (~100 atm).[25] A radical mechanism has been proposed (Scheme 21) in which CO is required to stabilize the cobalt carbonyl complex.

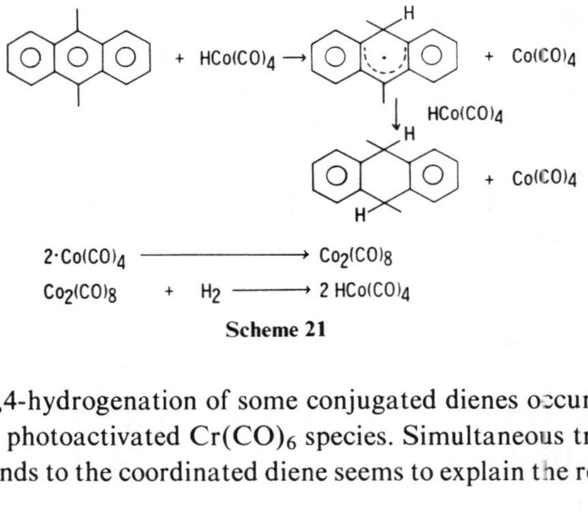

Scheme 21

Selective 1,4-hydrogenation of some conjugated dienes occurs with Cr(arene)(CO)$_3$ or photoactivated Cr(CO)$_6$ species. Simultaneous transfer of two hydrogen ligands to the coordinated diene seems to explain the remarkable selectivity[26]:

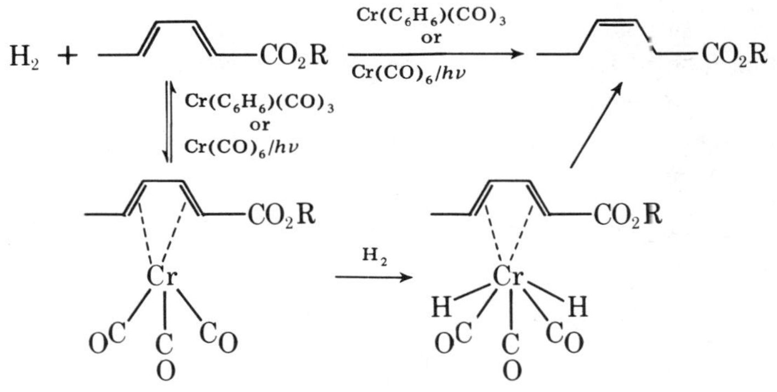

8. CARBONYLATIONS

Olefins or acetylenes react with carbon monoxide in the presence of relevant metal carbonyls as catalysts to give "carbonylation products." Carbonylation includes formation of aldehydes, ketones, higher alcohols, carboxylic acids, or their esters, depending on the coreactants (e.g., hydrogen, alcohols, or water) and on the reaction conditions. Four typical examples of carbonylation reactions appear below.[27]

1. Hydroformylation:

$$RCH=CH_2 + CO + H_2 \xrightarrow[\text{or}]{HCo(CO)_4} \begin{cases} RCH_2-CH_2CHO \\ RCH-CH_3 \\ | \\ CHO \end{cases}$$
$$ RhH(CO)(PR_3)_3$$

2. Hydrocarboxylation:

$$RCH=CH_2 + CO + H_2O \xrightarrow{Ni(CO)_4} \begin{cases} RCH_2-CH_2CO_2H \\ R-CH-CH_3 \\ | \\ CO_2H \end{cases}$$

3. Carbonylation of alcohol:
$$CH_3OH + CO \xrightarrow{[Rh(CO)_2I]_2/CH_3I} CH_3CO_2H$$

4. Hydroquinone formation:

$$2HC{\equiv}CH + CO \xrightarrow{Ru_3(CO)_{12}} \text{(hydroquinone)} \quad 63\%$$

The key step in these catalytic reactions is insertion of CO into a labile metal-carbon bond that is formed from a prior olefin- or acetylene insertion into a metal-hydrogen bond. The oxidative addition of alkyl halides also has been utilized in the formation of M—C bonds. General schemes for catalytic carbonylation have been established (Scheme 22).

The steps in the accepted mechanism for $HCo(CO)_4$-catalyzed hydroformylation are as follows[28a]:

1. *Dissociation* of CO to give a coordinatively unsaturated species (16-electron species):

$$HCo(CO)_4 \rightleftharpoons HCo(CO)_3 + CO$$

2. η^2-*Coordination* of the olefin to $HCo(CO)_3$:

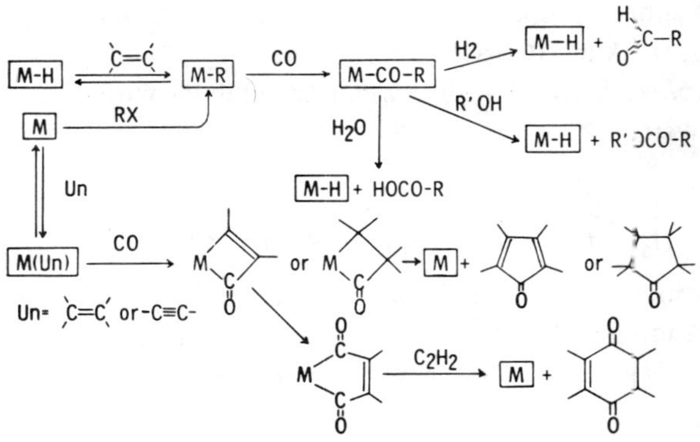

Carbonylations

Scheme 22. General scheme for catalytic carbonylation.

$$R—CH=CH_2 + HCo(CO)_3 \rightleftharpoons R—CH=CH_2 \\ \qquad\qquad\qquad\qquad\qquad\qquad | \\ \qquad\qquad\qquad\qquad\qquad HCo(CO)_3$$

3. *Insertion* of the coordinated olefin into the Co—H bond forming a labile alkylcobalt carbonyl. CO insertion then occurs by an attack of CO at the metal:

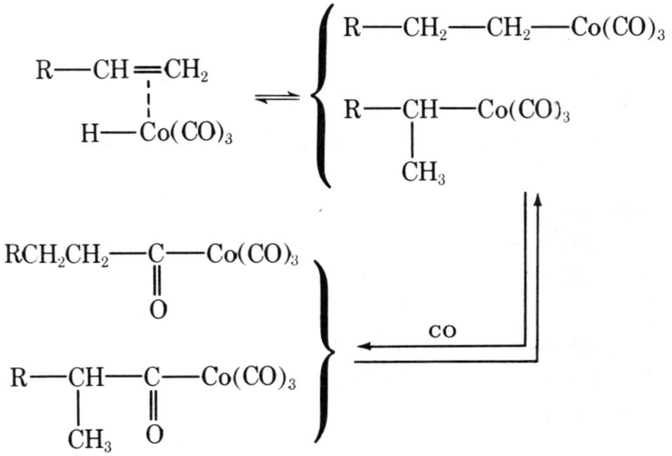

$$R—CH=CH_2 \atop H—Co(CO)_3 \rightleftharpoons \begin{cases} R—CH_2—CH_2—Co(CO)_3 \\ \\ R—CH—Co(CO)_3 \\ \quad\;\; | \\ \quad\;\; CH_3 \end{cases}$$

$$\begin{matrix} RCH_2CH_2—C—Co(CO)_3 \\ \parallel \\ O \\ \\ R—CH—C—Co(CO)_3 \\ \;\;|\quad\;\; \parallel \\ \;\;CH_3\;\; O \end{matrix} \Biggr\} \xleftarrow{CO}$$

4. *Oxidative addition* of H_2 to the coordinatively unsaturated acylcobalt species:

$$RCH_2—CH_2—\underset{\underset{O}{\parallel}}{C}—Co(CO)_3 + H_2 \rightleftharpoons RCH_2CH_2—\underset{\underset{O}{\parallel}}{C}—CoH_2(CO)_3$$

Recently it has been proposed that the cleavage of the cobalt-carbon bond is caused by an attack of $HCo(CO)_4$ rather than H_2.[28b]

5. *Irreversible reductive elimination* of an aldehyde with regeneration of the original catalyst:

$$RCH_2\text{---}CH_2\text{---}\overset{\overset{O}{\|}}{C}\text{---}CoH_2(CO)_3 \rightarrow RCH_2CH_2CHO + HCo(CO)_3$$

Carbonylation of conjugated dienes with Pd^{II} catalysts yields different products depending on the auxiliary ligand. Insertion of CO into a π-allyl palladium species is a key step for these reactions[29a]:

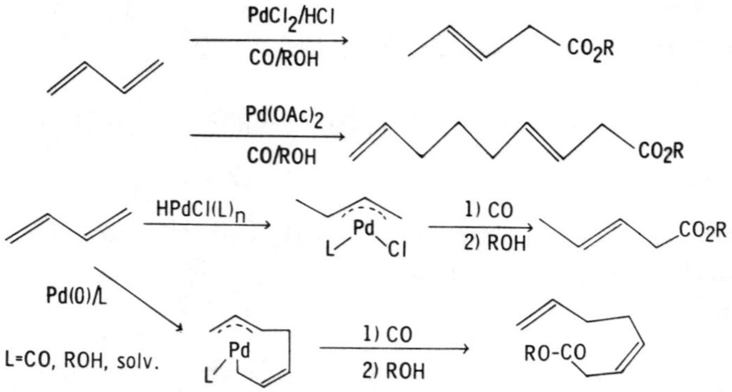

Carbonylation of methanol is an important method of production of acetic acid. Although this process has been investigated using cobalt carbonyl catalyst, the use of more expensive rhodium carbonyl catalyst systems such as $[RhI(CO)_2]_2/CH_3I/KI$ has led to high catalyst efficiencies coupled with low operating pressures and temperatures (170°). The mechanism involves (*a*) formation of methylrhodium species of oxidative addition of CH_3I to $[RhI_2(CO)_2]^-$, (*b*) CO insertion into the methyl-rhodium bond, (*c*) cleavage of the acetyl-rhodium bond by H_2O to give hydridorhodium species and acetic acid, (*d*) reductive elimination of HI from the hydridorhodium species to give $[RhI(CO)_2]_2$, and (*e*) formation of methyl iodide from HI and methanol[29b]:

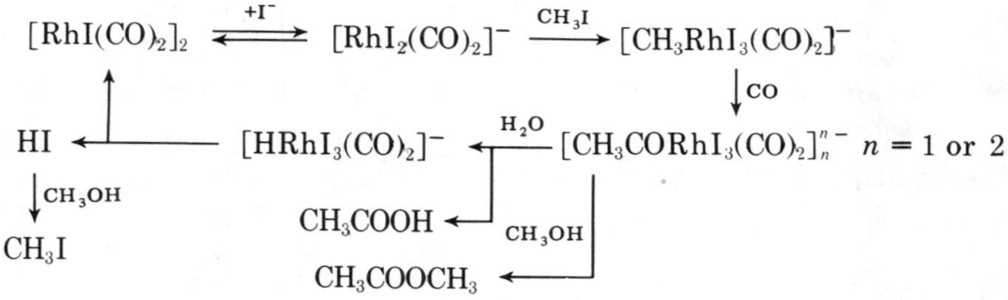

The rate of this whole catalytic reaction depends first order each in the concentrations of rhodium catalyst and CH_3I but independent in the concentrations of substrates CH_3OH and CO. This rate profile indicates a mechanism involving the rate-determining oxidative addition of CH_3I to $[RhI_2(CO)_2]^-$ (a square planar complex). Therefore the reaction can be performed at low carbon monoxide pressures and high conversion of CH_3OH.

9. OLIGOMERIZATIONS

Terminal olefins readily oligomerize under the influence of strong Lewis acids (e.g., $AlCl_3$) to a mixture of linear dimers, trimers, and higher oligomers with the composition being dependent on the reaction conditions. This oligomerization proceeds through a carbonium ion mechanism that is generally not selective to one particular oligomer:

$$CH_3CH{=}CH_2 \xrightleftharpoons{H^+} [CH_3{-}\overset{\oplus}{CH}{-}CH_3] \xrightarrow{CH_3CH=CH_2}$$

$$\begin{bmatrix} CH_3 & & CH_3 \\ \diagdown & & \diagup \\ CH{-}CH_2{-}CH \\ \overset{\oplus}{} & & \diagdown \\ & & CH_3 \end{bmatrix}$$

$$\updownarrow$$

$$\begin{bmatrix} & & & & CH_3 \\ & & & & \diagup \\ CH_3{-}CH_2{-}\overset{\oplus}{CH}{-}CH \\ & & & & \diagdown \\ & & & & CH_3 \end{bmatrix}$$

$$\updownarrow$$

$$\begin{bmatrix} CH_3 & CH_3 & CH_3 \\ | & | & \diagup \\ CH{-}CH_2{-}CH{-}CH_2{-}CH \\ \overset{\oplus}{} & & \diagdown \\ & & CH_3 \end{bmatrix} \xleftarrow{CH_3CH{=}CH_2}$$

$$\updownarrow$$

$$\begin{bmatrix} & & & & CH_3 \\ & & & & \diagup \\ CH_3{-}CH_2{-}CH_2{-}\overset{\oplus}{C} \\ & & & & \diagdown \\ & & & & CH_3 \end{bmatrix}$$

Isomeric carbonium ions

Selectivity in these olefin oligomerizations has been dramatically enhanced by utilizing certain organometallic complexes as homogeneous catalysts. For example, nickelocene or $(\eta^3$-allyl)$(\eta^5$-cyclopentadienyl)nickel acts as an effective catalyst for the selective linear dimerization of ethylene. The following paths have been suggested to explain the selectivity[30]:

Catalyst activation

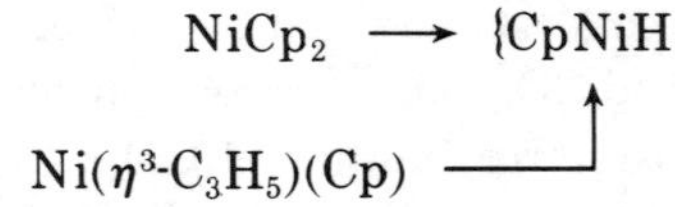

$$NiCp_2 \longrightarrow \{CpNiH\}$$
$$Ni(\eta^3\text{-}C_3H_5)(Cp) \longrightarrow$$

Catalytic cycle

$$\{CpNiH\} \underset{}{\overset{C_2H_4}{\rightleftarrows}} \left[\begin{array}{c} CpNi \leftarrow \overset{CH_2}{\underset{H}{\|}}_{CH_2} \end{array}\right] \rightleftarrows [CpNiC_2H_5]$$

$$CH_2=CHC_2H_5 \leftarrow$$

$$\left[\begin{array}{c} H \\ CpNi\cdots\overset{CH_2}{\underset{CH}{\|}} \\ C_5H_5 \end{array}\right] \rightleftarrows \left[\begin{array}{c} CH_2 \\ CpNi \overset{}{\underset{C_2H_5}{\diagdown}} CH_2 \end{array}\right] \rightleftarrows \left[\begin{array}{c} CH_2 \\ CpNi\cdots \overset{}{\underset{C_2H_5}{\diagdown}} CH_2 \end{array}\right] \Big\updownarrow C_2H_4$$

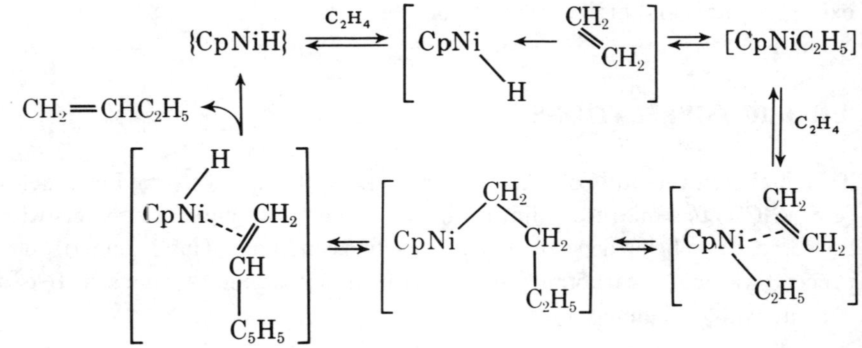

Numerous other combinations of nickel compounds with alkyl metals or hydrides have been also found to dimerize ethylene in the same way. The role of the η^5-C_5H_5 (i.e., Cp) ligand is to maintain and support the catalytically active nickel

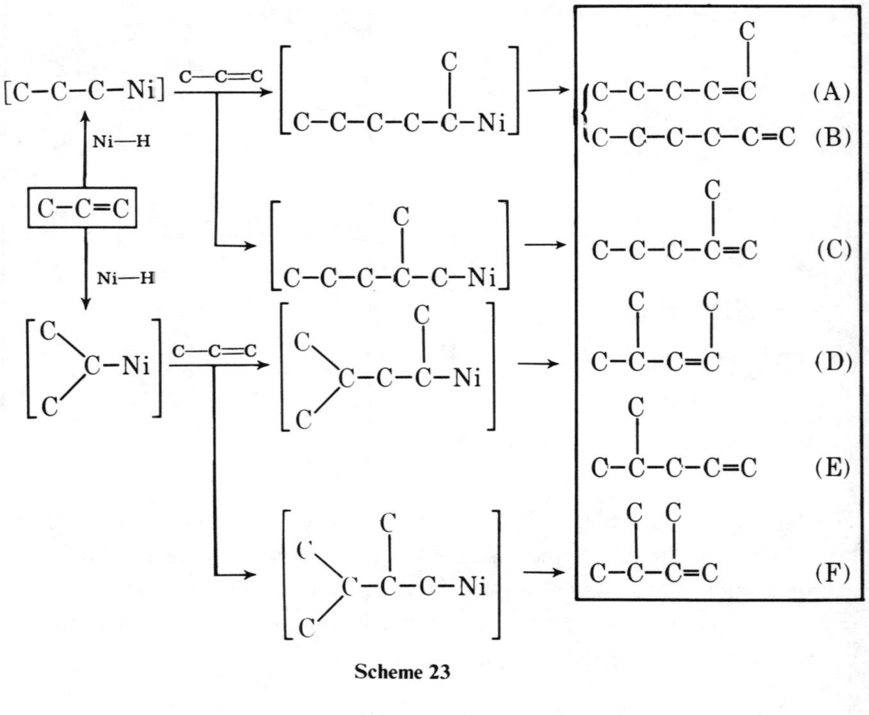

$$[C-C-C-Ni] \xrightarrow{C-C=C} \left[\begin{array}{c} C \\ | \\ C-C-C-C-C-Ni \end{array}\right] \longrightarrow \begin{cases} C-C-C-C=\overset{|}{C} \quad (A) \\ C-C-C-C-C=C \quad (B) \end{cases}$$

$$\Big\updownarrow Ni-H$$

$$\boxed{C-C=C}$$

$$\Big\downarrow Ni-H$$

$$\left[\begin{array}{c} C \\ \diagdown \\ C \diagup C-Ni \end{array}\right] \xrightarrow{C-C=C} \left[\begin{array}{c} C \\ \diagdown \\ C \diagup C-C-C-Ni \end{array}\right] \longrightarrow C-\overset{|}{\underset{}{C}}-C=\overset{|}{C} \quad (C)$$

$$\longrightarrow C-\overset{|}{\underset{}{C}}-\overset{|}{\underset{}{C}}=C \quad (D)$$

$$\longrightarrow C-\overset{|}{\underset{}{C}}-C-C=C \quad (E)$$

$$\left[\begin{array}{c} C \\ \diagdown \\ C \diagup C-\overset{|}{\underset{}{C}}-C-Ni \end{array}\right] \longrightarrow C-\overset{|}{\underset{}{C}}-\overset{|}{\underset{}{C}}=C \quad (F)$$

Scheme 23

Oligomerizations

species for many hundreds of catalytic cycles. In other words the catalyst life of the nickel species has been improved by the auxiliary Cp ligand. Linear dimerization of propylene with the $[(\eta^3\text{-}C_3H_5)NiCl]_2/Al_2Cl_3(Et)_3/2PR_3$ system produces a mixture of products.[31] Suitably substituted *tert*-phosphines remarkably enhance the product selectivity. Thus if $PCH_3\text{-}(1\text{-menthyl})_2$ is used as an auxiliary ligand, out of products A through F in Scheme 23, C is obtained with high selectivity (~84%). It has been suggested that these products are formed by way of insertion into the Ni—H bond, then insertion into a Ni—C bond, and finally β elimination.

Excellent selectivity for oligomerization of butadiene has been achieved with various types of nickel-ligand catalyst.[32] Scheme 24 gives examples.

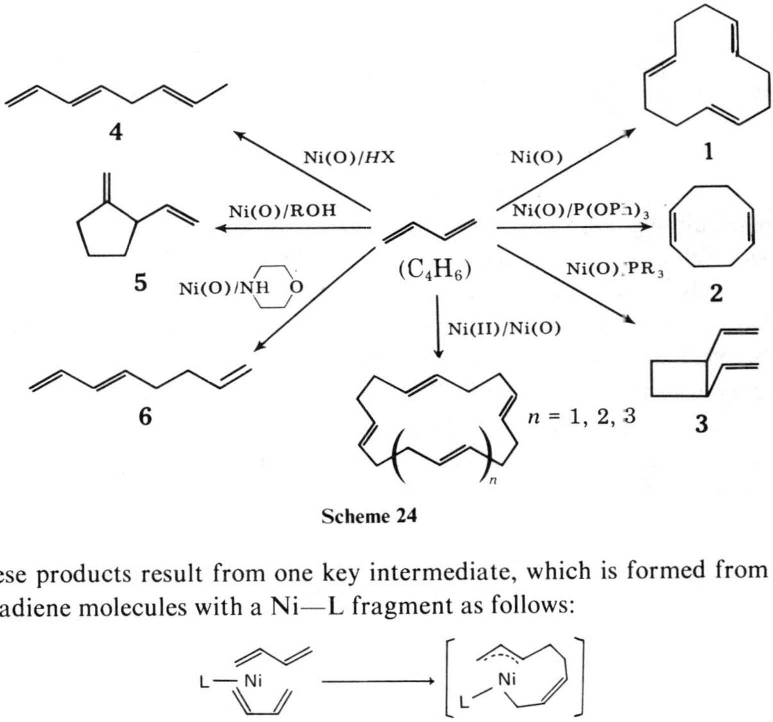

Scheme 24

These products result from one key intermediate, which is formed from two butadiene molecules with a Ni—L fragment as follows:

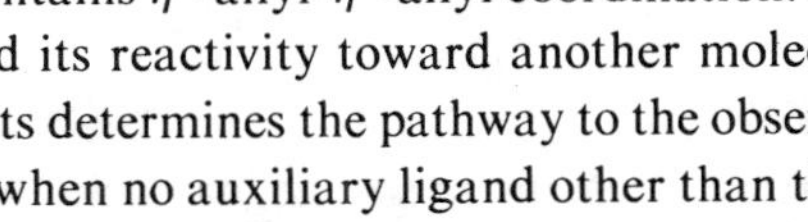

This intermediate contains η^3-allyl-η^1-allyl coordination. A combination of its thermal stability and its reactivity toward another molecule of butadiene or toward protic reagents determines the pathway to the observed products. When L = C_4H_6—that is, when no auxiliary ligand other than the reactant itself exists—insertion of the coordinated C_4H_6 occurs, resulting in the formation of a bis-η^3-allyl-type intermediate that yields the cyclic trimer. Triarylphosphine

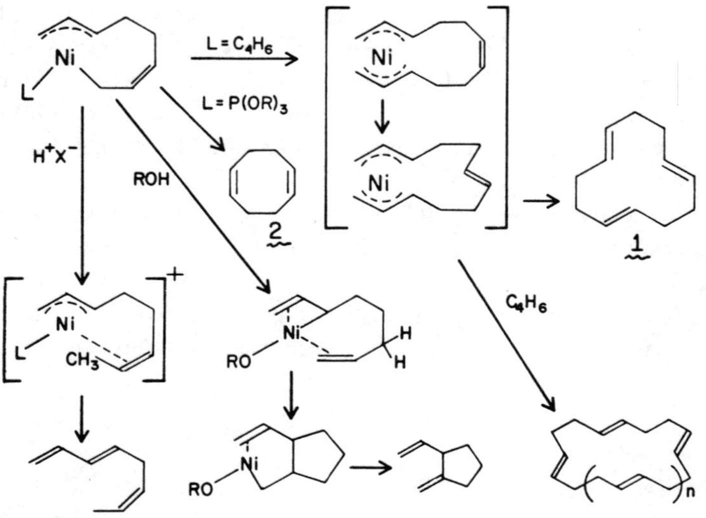

Fig. 53. Mechanisms for the cyclooligomerization of butadiene.

or -phosphite ligands prevent the formation of the C_4H_6 trimer and facilitate reductive elimination to cyclic dimers **2** and **3** in Scheme 24. Compound **3** has been readily transformed to **2** by the same catalyst (see Fig. 53). Co-oligomerization of C_2H_4 and C_4H_6 occurs with the same catalyst system.[33] This reaction has been extended to involve many other unsaturated organic compounds, such as acetylenes (RC≡CR) and allenes (CH_2=C=CR_2). When L is tri(o-phenylphenyl)phosphite, the cyclooligomers are the major products. Trialkylphosphines change the course of the reaction that results in the linear co-oligomer:

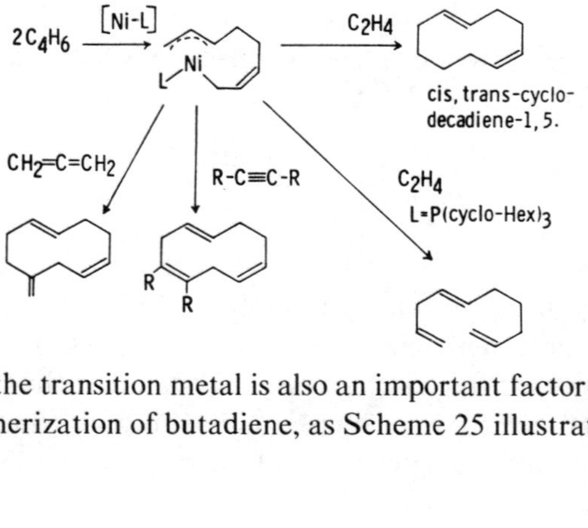

The identity of the transition metal is also an important factor determining the course of oligomerization of butadiene, as Scheme 25 illustrates.

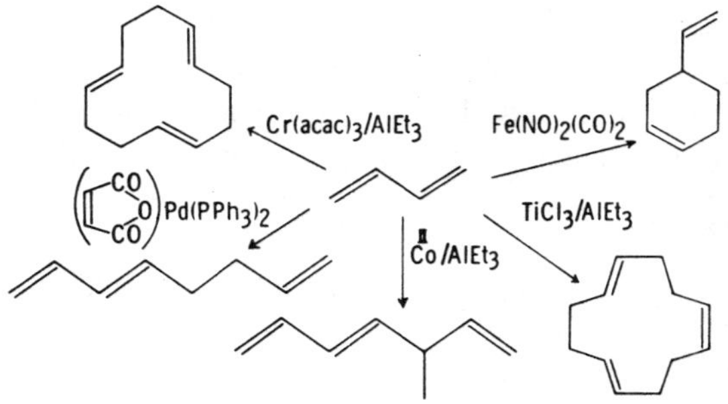

Oligomerizations

Scheme 25

Cyclooligomerization of acetylene to benzene, cyclooctatetraene, and so on, occurs in the presence of suitable nickel or cobalt complexes [e.g., $Ni(acac)_2$]. Although the catalytically active species has not been isolated, metal-acetylene and metallocyclopentadiene complexes have been thought to be the key intermediates[34]:

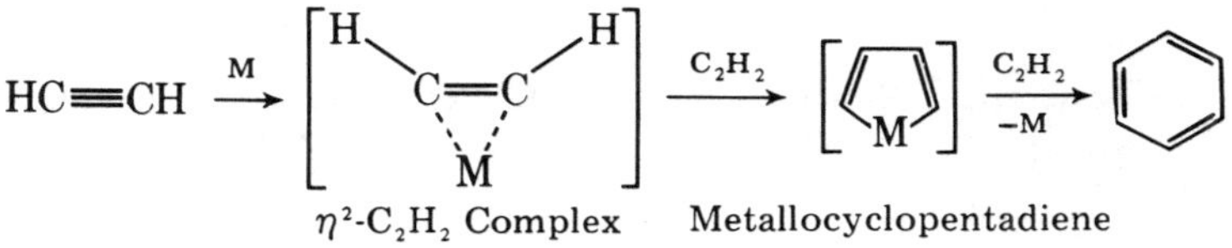

$$HC\equiv CH \xrightarrow{M} \left[\begin{array}{c} H\diagdown\diagup H \\ C=C \\ \diagdown\diagup \\ M \end{array}\right] \xrightarrow{C_2H_2} \left[\begin{array}{c}\\ M \end{array}\right] \xrightarrow[-M]{C_2H_2} \bigcirc$$

η^2-C_2H_2 Complex   Metallocyclopentadiene

Zerovalent nickel complexes [e.g., $Ni(\eta^3$-$C_3H_5)_2$] lacking a strongly coordinating ligand, such as PPh_3, catalyze the formation of cyclooctatetraene, which may be formed by way of a metallocycloheptatriene intermediate. The formation of a small amount of naphthalene, azulene, and vinylcyclooctatetraene implies that nine-membered ring, $\underline{M-(CH=CH)_4}$ is also an intermediate in the catalytic cycle.

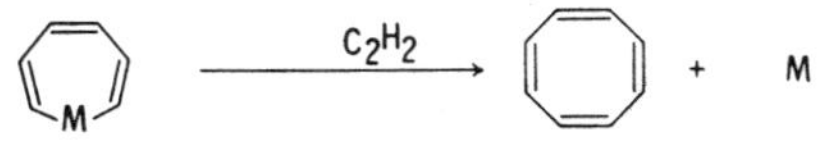

Linear oligomerization of C_2H_2 to mono- and divinylacetylene occurs with Niewland's catalyst ($CuCl/NH_4Cl/aq$ HCl). Unstable copper acetylide seems to be the active intermediate that allows another C_2H_2 molecule to insert in the Cu—C bond:

$$[Cu-C\equiv CH] \xrightarrow{C_2H_2} [Cu-CH=CH-C\equiv CH] \xrightarrow{HCl}$$

$$CuCl + CH_2=CH-C\equiv CH$$

NH_4Cl is necessary for the stabilization of the active catalyst in aqueous solution.

Some organopalladium complexes are active for the selective linear trimerization of phenylacetylene.[35a] A similar C≡C insertion into the Pd—C bond has been suggested:

$$PhC\equiv CH \xrightarrow{PdPh_2(PEt_3)_2} Ph-CH=C(Ph)-CH=CH-Ph$$

Dimerization of a nickelacyclopentadiene species has been proposed as an intermediate reaction in the Ni-complex-catalyzed cyclotetramerization of acetylene to cyclooctatetraene[35b]:

$$HC\equiv CH \rightarrow [\text{nickelacyclopentadiene}] \rightarrow [\text{Ni-Ni bridged dimer}] \rightarrow [\text{cyclooctatetraene}]$$

The cyclooligomerization of acetylene with olefins, nitriles, or isocyanate has been catalyzed by $CoCp_2$, $CoCp(1,5\text{-}COD)$, or $Co(\eta^3\text{-}C_8H_9)(1,5\text{-}COD)$[36]:

$$2CH\equiv CH + M \rightarrow [\text{metallacyclopentadiene, M}]$$

with:
- RCN → 2-R-pyridine
- $EtO_2CCH=CH-CO_2Et$ → cyclohexadiene-1,2-dicarboxylate (CO_2Et, CO_2Et)
- R—N=C=O → N-R-2-pyridone

M = CoCp

A metallocyclopentadiene has been confirmed as the key intermediate in these cyclo co-oligomerizations.

Cyclooligomerization of allene to four, six, eight, or ten-membered ring exomethylene compounds proceeds selectively using a Ni(0) catalyst with the proper auxiliary ligands and reaction conditions:

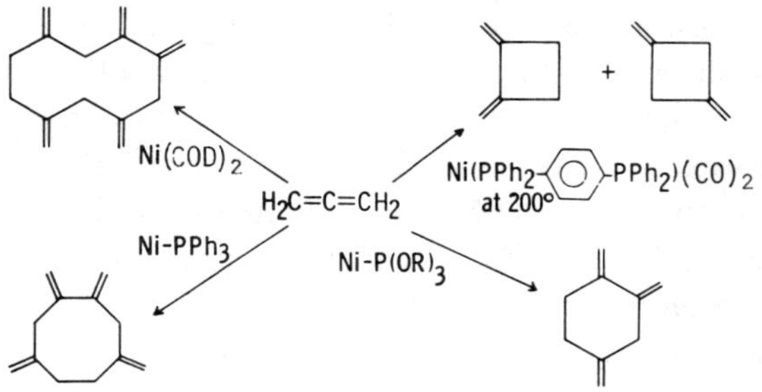

Nickel(0) species with chelating bisallyl ligands have been the common intermediates utilized in these selective reactions[34]:

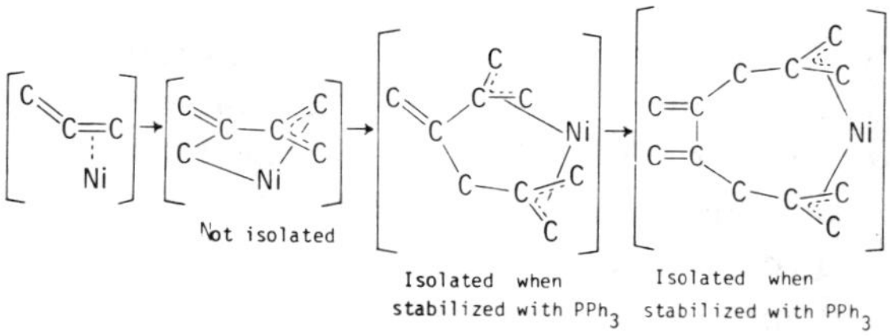

10. POLYMERIZATIONS

Homogeneous catalysts have been used very advantageously for polymerizations in the laboratory and also in industry.[37] Ionic polymerizations have been initiated by a variety of electrophilic or nucleophilic catalysts, so-called initiators. Electron-rich olefins such as isobutylene have been polymerized by a typical electrophile (σ acceptor) such as BF_3 at low temperature. The carbonium cation

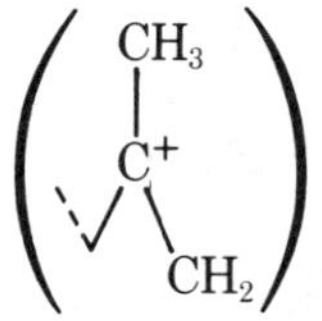

is the active propagating species in this polymerization, which had been generally called cationic polymerization.

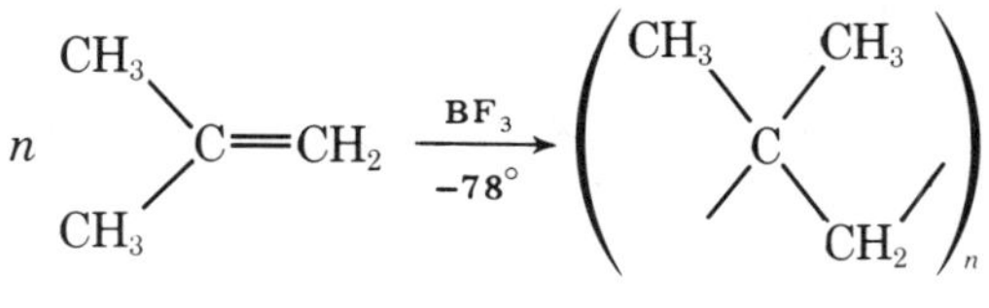

Olefins with one or more electron-withdrawing substituents have been polymerized by anionic or nucleophilic initiators. Strongly polarized and electron-deficient double bonds (e.g., $CH_2=C(CN)_2$) are activated toward polymerization even by weakly nucleophilic reagents such as water, sodium hydroxide, and amines. The anionic polymerization initiated by polar metal alkyls is influenced by the metal ion (cationic counterion) because the metal ion is always situated near the propagation site. Thus alkyllithium can produce only 1,4-*cis*-polyisoprene, but alkylsodium or dialkylmagnesium yields a sterically uncontrolled polymer:

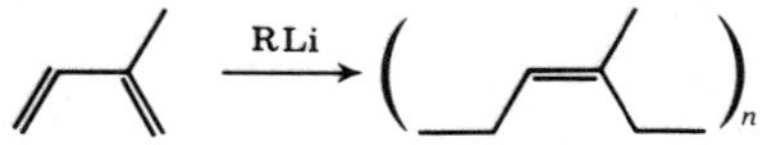

$$\text{isoprene} \xrightarrow{\text{RLi}} (\text{polyisoprene})_n$$

Radical polymerization initiators have often been regarded as "catalysts." Here the free radical, generated by thermal or photochemical decomposition of organic azo or peroxy compounds, adds to the double bond of an olefin to give a propagating radical end (Schemes 26 and 27). The same propagating radical can also be generated by an electron transfer redox reaction involving low-valent transition metal compounds and polyhaloalkanes.

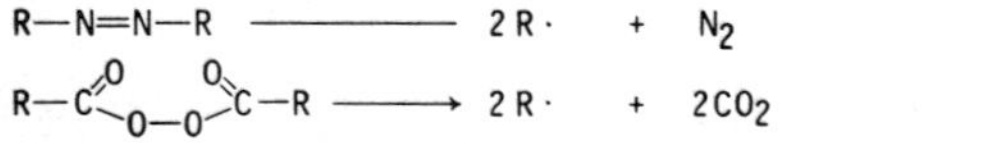

$$R-N=N-R \longrightarrow 2R\cdot + N_2$$
$$R-\overset{O}{C}-O-O-\overset{O}{C}-R \longrightarrow 2R\cdot + 2CO_2$$
$$M(CO)_n + CCl_4 \longrightarrow [M(CO)_n]^+[CCl_4^{\cdot-}] \longrightarrow \cdot CCl_3 + Cl^-$$

Scheme 26. Formation of free radicals.

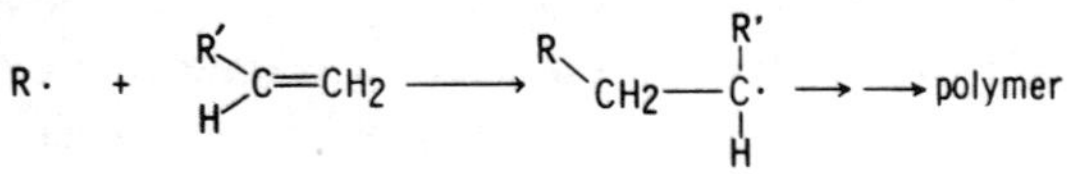

$$R\cdot + \underset{H}{\overset{R'}{C}}=CH_2 \longrightarrow R-CH_2-\underset{H}{\overset{R'}{\underset{|}{C}}}\cdot \longrightarrow \longrightarrow \text{polymer}$$

Scheme 27. Propagation in radical-chain polymerization.

In the presence of a large excess of radical-chain transfer compounds, propagation does not last long, and the radical-initiated reaction turns out to give telomers. The propagating radical abstracts a chlorine atom from CCl_4 and generates $\cdot CCl_3$, which adds to the double bond to give another radical:

$$PhCH=CH_2 + CCl_4 \xrightarrow{H_2Ru(PPh_3)_4}$$

$$Ph-\underset{\bullet}{C}H-CH_2-CCl_3 \xrightarrow{CCl_4} Ph-\underset{\underset{Cl}{|}}{C}H-CH_2-CCl_3 + \cdot CCl_3$$

Stereoregular polymerization of olefins and dienes is one of the best examples of the excellence of homogeneous metal combination catalyst systems (e.g., Ziegler catalysts). The original Ziegler combination catalyst, $TiCl_4/AlEt_3$ in heptane, has both soluble and insoluble parts that are active for the high polymerization of ethylene. For *isotactic* polymerization of propylene, only the insoluble part has catalytic activity. The heterogeneous catalyst prepared from solid $TiCl_3$ and $AlEt_3$ is a far better one for this purpose:

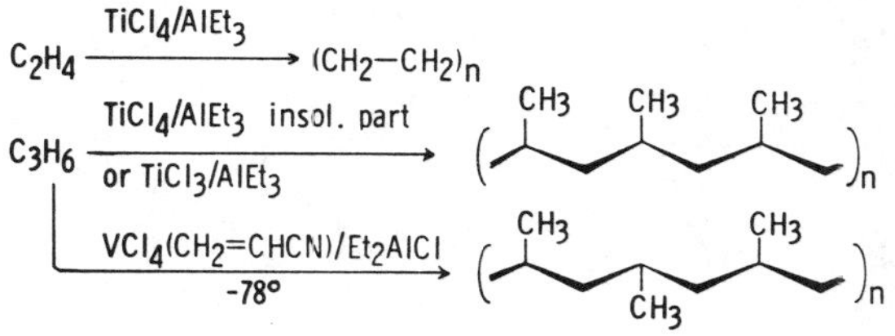

$$C_2H_4 \xrightarrow{TiCl_4/AlEt_3} (CH_2-CH_2)_n$$

$$C_3H_6 \xrightarrow[\text{or } TiCl_3/AlEt_3]{TiCl_4/AlEt_3 \text{ insol. part}} \left(\begin{array}{c} CH_3 \quad CH_3 \quad CH_3 \\ \diagdown\diagup\diagdown\diagup \end{array} \right)_n$$

$$\xrightarrow[-78°]{VCl_4(CH_2=CHCN)/Et_2AlCl} \left(\begin{array}{c} CH_3 \quad\quad CH_3 \\ \diagdown\diagup\diagdown\diagup \\ CH_3 \end{array} \right)_n$$

A homogeneous catalyst $[VCl_4 \cdot (CH_2=CHCN)/Et_2AlCl]$ for propylene polymerization yields syndiotactic polypropylene at $-78°$, whereas another complex $VO(OR)_3/Et_2AlCl$, produces only atactic polypropylene. Titanium-based homogeneous systems such as $TiCl_2(\eta\text{-}C_5H_5)_2/AlEt_3$ are generally less catalytically active for C_2H_4 and C_3H_6.[38] In spite of these apparent differences, the essential steps in the polymerization are quite similar in both homogeneous and heterogeneous systems. The homogeneous systems are as complex as the heterogeneous ones. Since its discovery in 1956, the Ziegler catalyst system has been critically examined, and the coexistence of various kinds of reactive species has been found [e.g., Ti(II) and Ti(III) metal species]; some of the species are coordinated by anionic ligands such as Cl^- and OR^-, and some are bridged by alkylaluminum species. The Lewis acidity and steric bulk of the aluminum component are properties that affect the Ti center indirectly by way of donor-acceptor interaction through the anionic ligands:

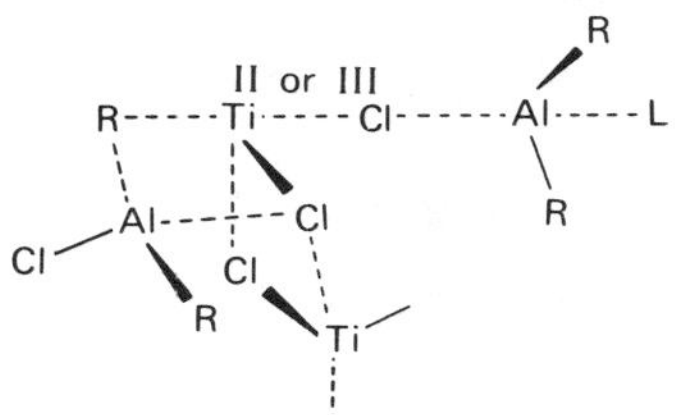

Small amounts of protic reagents, Lewis acids (electrophiles), or bases (nucleophiles), often activate or deactivate the catalytically active centers. The complexity and delicacy of the Ziegler catalysts have been revealed by kinetics, by examination of the polymers formed, and by other physical techniques. A detailed mechanism of olefin polymerization by *soluble* Ziegler catalysts remains unknown. In contrast, the active site structure in a heterogeneous Ziegler-Natta catalyst $TiCl_3/AlEt_3$ has been best described by the well-known Cossee mechanism, which involves selective insertion of an olefin (e.g., propylene) into a reactive Ti^{III}—C bond situated at the surface of α-$TiCl_3$:

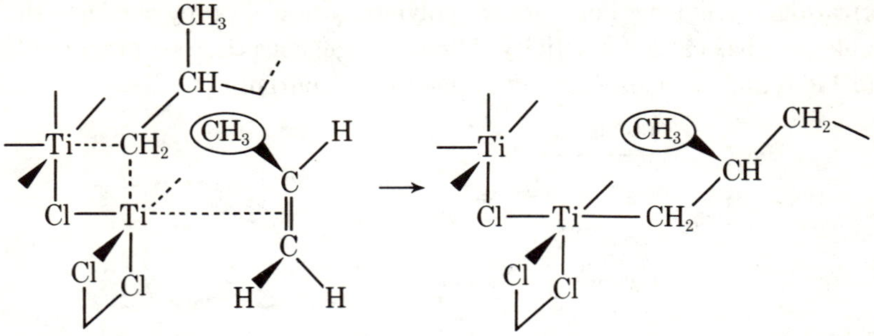

Further support of the importance of a coordinatively unsaturated (or free-valence state) Ti^{III} site has been furnished by means of an electron microscope, which permitted the observation of the growth of a polymer at the *edge* of a $TiCl_3$ crystal. Soluble single-metal catalysts for polymerization of ethylene and propylene are known. Tetrabenzyl compounds of titanium and zirconium are thermally stable because β hydrogens are absent from the alkyl group. These relatively stable alkyls form homogeneous solutions in nonpolar inert solvents and catalyze apparently Ziegler-type polymerizations without the addition of alkyls of metals in Groups I to III.[39] Some interaction of the aromatic π electrons of the benzyl group seems to help the stability of the active species against a homolytic cleavage of the metal-carbon bonds:

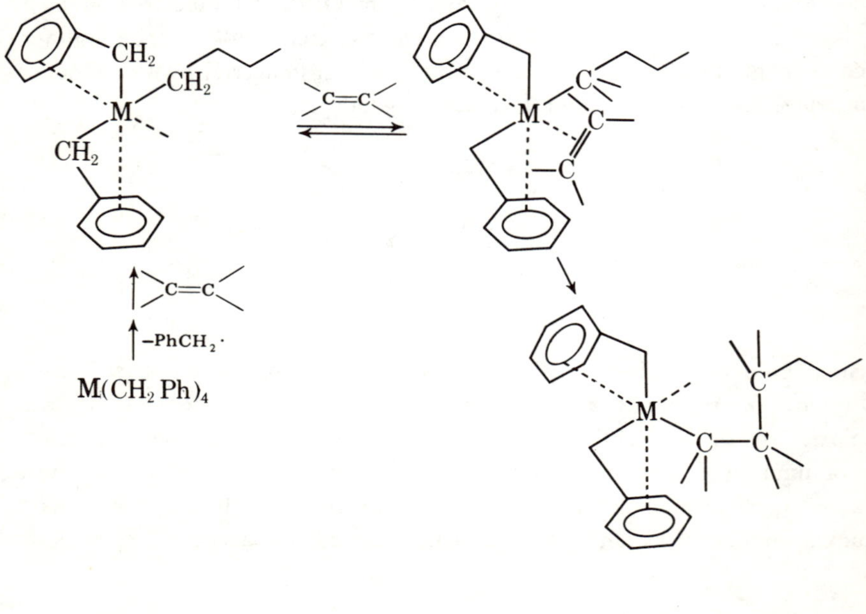

Polymerizations

Obviously the insertion of the terminal olefin into a reactive metal-carbon bond on the single-metal active species is the most important propagation step. The apparent homogeneity of the catalyst, however, does not necessarily indicate that the active site is homogeneous. The well-known Ziegler system $VCl_4/AlEt_2Cl$ in toluene, which is apparently homogeneous, polymerizes propylene at $-21°$ to $10°$ to an *isotatic polypropylene,* and careful examination of this compound revealed that the polymer must be formed from a small amount of heterogeneous active sites. At temperatures lower than $-21°$, the same homogeneous catalyst gives syndiotactic polymers.[40a] Recently it has been proposed that metal-carbene species constitute the active catalyst in stereoregular polymerization.[40b]

Butadiene has been polymerized by the Ziegler-type combination systems. Table 15 gives typical examples of how selectivity has been influenced by the identity of transition metal. No simple theory can explain these differences in selectivity. However it is now evident that a η^3-allyl-metal complex is the active species, and its stereochemistry in the reaction with attacking butadiene determines the product (see Scheme 28).

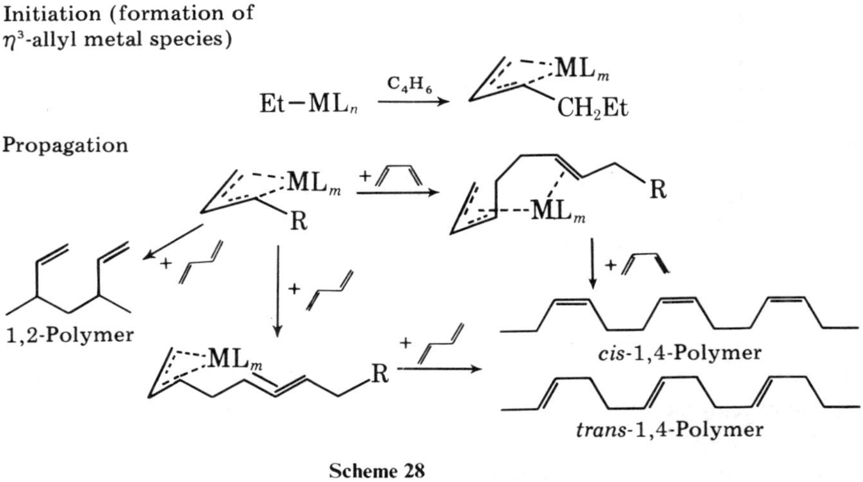

Initiation (formation of η^3-allyl metal species)

$$Et-ML_n \xrightarrow{C_4H_6} \cdots ML_m / CH_2Et$$

Propagation

Scheme 28

A suitably chosen catalyst has yielded a 1-1 stereochemical mixture of polybutadienes (called an equibinary polymer)[41a,b]:

$$C_4H_6 \rightarrow \text{poly-}cis,trans\text{-1,4-Butadiene}$$

Table 15. Stereoselective Polymerization of Butadiene

	Microstructure of the Polymer (%)		
	trans-1,4	cis-1,4-	1,2-
$CoCl_2/AlEt_2Cl$	1	98	1
$MoO_2(OR)_2/AlEt_3$	1	3–6	92–96 (syndiotactic)
$Cr(acac)_3/AlEt_3$	1–2	0–3	97–99 (isotactic)
$VCl_3(THF)_3/AlEt_2Cl$	99	0	1

S. M. Atlas and H. F. Mark, *Cat. Rev.*, **13**, 1 (1976).

Stereoregular crystalline *cis*-polyacetylene [poly(vinylene)] has been prepared by contact of gaseous acetylene to the surface of a solution of a homogeneous catalyst prepared by mixing $AlEt_3$ and $Ti(O\text{-}n\text{-}Bu)_4$:

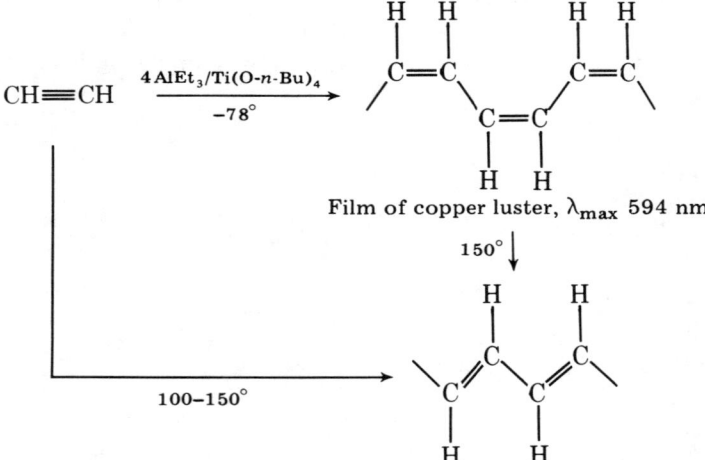

The same catalyst system has also produced the corresponding *trans*-polymer at higher temperature. These polymers have a metallic luster and are semiconductors (10^{-5} to $^{-9}$ Ω^{-1} cm^{-1}). The doping of electron acceptors or donors (e.g., AsF_5 or Na) has remarkably enhanced the electric conductivity up to a factor of 10^{12}. A film of composition $CH(AsF_5)_{0.064}$ exhibits a conductivity of 10^3 Ω^{-1} cm^{-1}. Applications of these compounds to electronic devices are now being actively investigated.[41c-g]

Highly regular head-to-tail polymers of allene have been prepared by passing allene gas into an ethanol solution of $[RhCl(CO)_2]_2$. Addition of two molar equivalents of Ph_3P accelerates the rate of polymerization. The polymer has a high melting point (124°) and is quite stable in air for a long time (5 years) in spite of the presence of a large number of terminal methylene groups:

$$CH_2=C=CH_2 \xrightarrow[\text{EtOH at 20°}]{[RhCl(CO)_2]_2/2\ Ph_3P}$$

[structure: polymer with repeating $C(=CH_2)$ units connected via CH and CH_2]

Colorless crystalline polymer insoluble in organic solvents

The polymer is insoluble in most organic solvents at room temperature because of its high crystallinity. Conventional polymerization initiators do not polymerize allene to high regular polymers. Therefore homogeneous catalysts have a distinct advantage over other catalyst systems in producing these insoluble polymers free from the contamination of the inactive catalyst. The catalytic activity for the allene polymerization has been also found for some other transition metal complexes. For example, $CoCl(Ph_3P)$ suspended in a two-phase system of benzene and water containing micellar agents rapidly polymerizes allene to a crystalline polymer at 5°.[41h-g]

η^3-Allyl-metal compounds are active catalytic species for allene polymerization as evidenced by their stoichiometric reactions; for example, an Me-Pd compound inserts allene to give a

$$\eta^3\text{-}(CH_3\text{---}C\begin{smallmatrix}CH_2\\CH_2\end{smallmatrix})Pd \text{ species}$$

11. OXIDATIONS

Various types of oxidation have been catalyzed homogeneously by transition metal complexes.[42,43] One of the most important examples is oxidation of olefins by the Wacker process. An aqueous hydrochloric acid solution of a $PdCl_2/CuCl_2$ system is effective in converting ethylene to acetaldehyde by air oxidation. The proposed mechanism involves η^2 coordination of C_2H_4 to Pd(II), nucleophilic attack of OH^- on the η^2-C_2H_4, and β-hydrogen transfer to Pd(II) with liberation of acetaldehyde or formation of a labile vinylalcohol complex. These processes produce metallic palladium, which is then oxidized by an effective oxidation system, O_2/CuCl/aq HCl, to $PdCl_2$. (see p. 93 and 102).

1. C_2H_4 Oxidation:

$$C_2H_4 \underset{PdCl_2/HCl}{\rightleftharpoons} \begin{bmatrix} CH_2 \\ \| \text{----} PdCl_3 \\ CH_2 \end{bmatrix}^- \underset{H_2O}{\rightleftharpoons} \begin{bmatrix} OH_2 \\ CH_2 \; | \\ \| \text{----} Pd\text{---}Cl \\ CH_2 \; | \\ \quad Cl \end{bmatrix}$$

$$\begin{Bmatrix} CH_3CHO \\ Pd \; + \; HCl \end{Bmatrix} \longleftarrow \begin{bmatrix} CH_2\text{---}OH \\ | \\ CH_2\text{---}PdCl_2 \end{bmatrix}^- \underset{-H^+ \; | \; H_2O}{\longleftarrow} $$
$$(OH_2)$$

(fast) $\nwarrow$ $\swarrow$

$$[HPd(CH_2\text{=}CHOH)Cl]$$

2. Metal oxidation:

$$Pd + 2CuCl_2 \longrightarrow PdCl_2 + 2CuCl$$
$$2CuCl + O_2 + 2HCl \longrightarrow 2CuCl_2 + H_2O$$

This oxidation has not been limited to C_2H_4, but is applicable to other olefins as well. Propylene and 1-butene have been converted to acetone and methylethylketone, respectively. In the presence of acetic acid, the oxidation catalysis of ethylene yields vinyl acetate with the nucleophilic attack of acetate anion being involved in the key catalytic step:

$$CH_2\text{=}CH_2 + CH_3CO_2H \xrightarrow[O_2/Cu(OAc)_2]{Pd(OAc)_2} CH_2\text{=}CH\text{---}OAc$$

Similar acetoxylation has occurred when toluene or xylene has been used. The observed activation of a C—H bond by palladium catalysis has been induced by coordination of the aromatic ring to Pd(II). Similarly, orthopalladation of some aromatic compounds occurs through coordination of Pd(II) to the amine, ether, or C=C group of the aromatic side chain. Dehydrogenative dimerization of some olefins or aromatics occurs when $Pd(OAc)_2$ has been used in a stoichiometric amount:

$$Ph\text{-}CH_3 + AcOH \xrightarrow[\text{air}]{Pd(OAc)_2} Ph\text{-}CH_2OAc$$

$$Ph\text{-}[GR] \xrightarrow{PdX_2} Ph\text{-}[GR]\text{---}PdX_2 \longrightarrow Ph\text{-}[GR]\text{---}PdX$$

$$[GR] = -O-, \; -N\!\!\diagup\!\!\diagdown, \; \diagdown\!\!C\text{=}C\!\!\diagup$$

$$2 \underset{R'}{\overset{R}{>}}C=CH_2 + Pd(OAc)_2 \xrightarrow{80-100°}$$

$$\underset{R'}{\overset{R}{>}}C=CH-CH=C\underset{R'}{\overset{R}{<}} + Pd + AcOH$$

[benzene-R] + Pd(OAc)$_2$ $\xrightarrow{\text{in AcOH}}$ [R-biphenyl-R]

Mixed dehydrogenative coupling between aromatics and olefins also occurs when a stoichiometric amount of $Pd(OAc)_2$ is used. In the course of the reaction, metallic palladium separates and can be recycled by air oxidation in separate reactions:

$$CH_2=CH_2 + [benzene] \xrightarrow[AcOH]{Pd(OAc)_2} [Ph-CH=CH_2] + Pd$$

$$\downarrow Pd(OAc)_2, AcOH, [benzene]$$

$$[Ph-CH=CH-Ph] + Pd$$

A mechanism involving the initial substitution of aromatic hydrogen with Pd and the insertion of C_2H_4 into the resulting Pd—C bond has been proposed to explain the observed selectivity in the mixed coupling reaction.[44]

Allylic oxidation of olefins occurs stoichiometrically with Hg(II), Tl(III), or Pd(II) in acidic media. When coupled with a metal oxidation-reduction process, these stoichiometric reactions have been made catalytic.

Air oxidation of cumene has been catalyzed by Co(II) or Mn(II) salts of higher carboxylic acids (e.g., laurate).[45] Since the oxidation proceeds by a radical-chain mechanism, the reaction usually occurs after some induction period; that is, the initiation step is slower than the main chain reaction. This characteristic tends to make the observer think that the reaction has occurred automatically just because the reaction has been allowed to stand. Therefore this has been called "autoxidation."

A one-electron transfer from an olefin to Co^{III}, Mn^{III}, Pb^{IV}, or Ce^{IV} has been considered to be an important initiation reaction of these autoxidations. For example, $[Co^{III}(R'CO_2)_2]^+$ works as a trigger to generate a very reactive radical cation from α-olefins, as follows:

$$RCH=CH_2 + [Co^{III}(R'CO_2)_2]^+ \longrightarrow (R\overset{+}{C}H-\overset{\bullet}{C}H) + Co^{II}(R'CO_2)_2$$

$$(R\overset{+}{C}H-\overset{\bullet}{C}H_2) + O_2 \rightleftarrows (R\overset{+}{C}H-CH_2OO\bullet)$$

$$PhCH(CH_3)_2 \longrightarrow Ph\overset{\bullet}{C}(CH_3)_2 \xrightarrow{+O_2} \underset{\underset{OO\bullet}{|}}{PhC(CH_3)_2} \longrightarrow \underset{\underset{OOH}{|}}{PhC(CH_3)_2}$$

Phenol and catechol have been oxidized to monomethyl muconate by an O_2/CuCl/py system in methanol:

[Structural diagram: phenol or catechol → (O$_2$, MeOH) → muconate diester with CO$_2$CH$_3$ and CO$_2$H groups]

It has been proposed that a Cu(II) species formed by air oxidation is the active species.[46]

Biochemical oxidations utilize the redox properties of transition metals in special environments. Electron transfer between substrate and active metal sites containing copper, cobalt, iron, and so on, is the most important portion of many biochemical mechanisms. Chemical activation of O_2 by a synthetic Co^{II} complex has been found to take place as follows:

[Structural diagram: Co(bzacen) + O_2 →(py) Co^{III}–O$_2^{\bullet-}$ complex with pyridine coordinated]

Co(bzacen)

The reactions in Scheme 29 represent synthetic models of quercetinase and tryptophan pyrolase. Activation may be due to one-electron transfer from the Co^{II} to O_2 because the Co—O_2 complex can be prepared by the reaction of O_2^- with the corresponding Co^{III}.[47]

Model compounds of Cu^{II} have been found to possess similar catalytic activity

Oxidations

to those catalyzed by Co(Salen) but in lower efficiency. Also an organometallic Co complex CpCo^{I}C$_5$H$_5$—O—O—C$_5$H$_5$Co^{I}Cp, prepared from the reaction of Cp$_2$Co and O$_2$, exhibits very similar oxidations, such as the oxidative cleavage of α-diketones and o-quinones yielding cobalticinium carboxylates.[49]

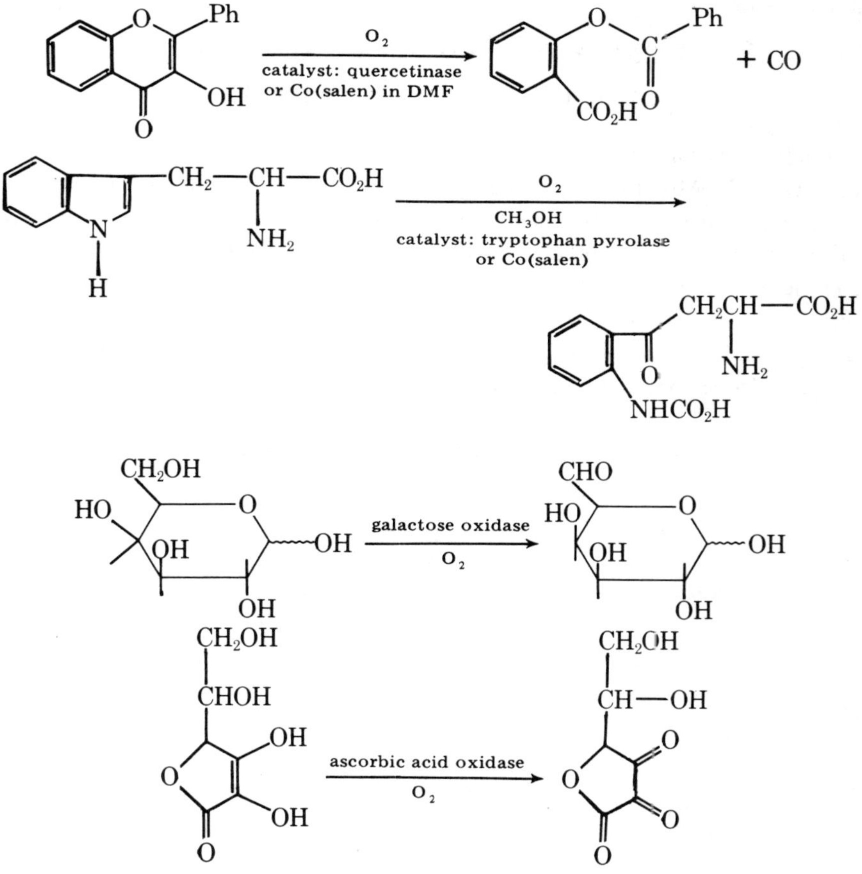

Scheme 29

The activation of oxygen by formation of "side-on" dioxygen metal complexes (η^2-coordinated O$_2$) for air oxidation of *tert*-phosphines or alkylisocyanides is also known. A stepwise transfer of O$_2$ has been suggested as a mechanism.[50a] Paths involving HO$_2^-$ have also been proposed[50b]:

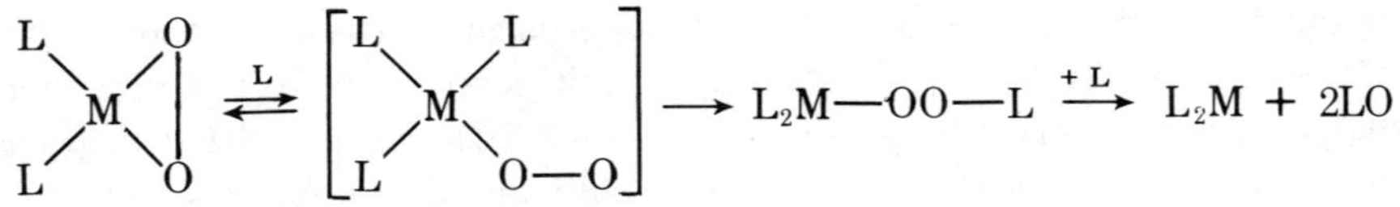

$$L\diagup_L^{\diagdown}M\diagup_O^{\diagdown}O \;\rightleftarrows\; \left[L\diagup_L^{\diagdown}M\diagup_{O-O}^{\diagdown L}\right] \longrightarrow L_2M{-}OO{-}L \xrightarrow{+L} L_2M + 2LO$$

For example,

$$2RNC + O_2 \xrightarrow{Ni(RNC)_4} 2R{-}N{=}C{=}O$$

$$2Ar_3P + O_2 \xrightarrow{M(PAr_3)_4} 2Ar_3P{=}O$$

Olefin epoxidation with hydrogen peroxide has been catalyzed by Na_2WO_4 or Na_2MoO_4. Modification of these catalysts utilizing suitable ligands extends the applicability and efficiency of the epoxidation. Thus N,N-dimethylformamide (DMF) has been used as the ligand for $MoO_5(DMF)$, which is generally soluble in an olefin/EtOH/H_2O reaction mixture and is an active catalyst at 30° at concentrations as low as 0.1 mole %. Other peroxymolybdenum or peroxytungsten compounds have also been found to be catalytically active.[43] Organic hydroperoxides are also used as oxidants:

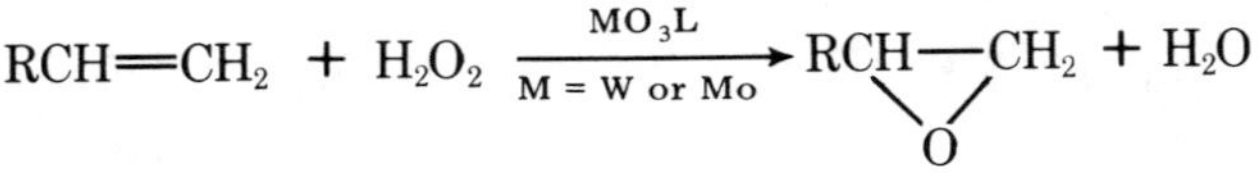

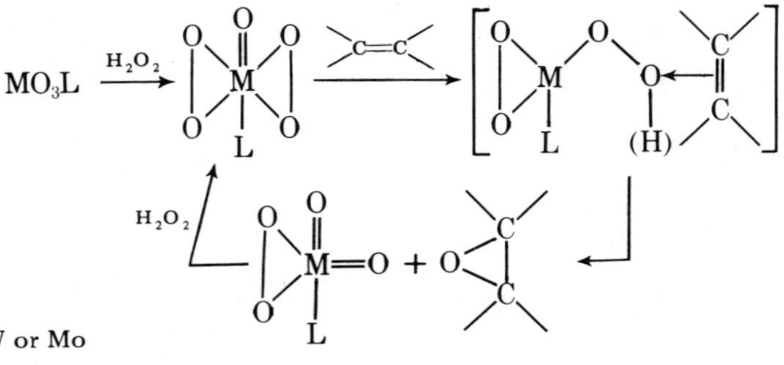

$$RCH{=}CH_2 + H_2O_2 \xrightarrow[M = W \text{ or } Mo]{MO_3L} RCH{-}CH_2 + H_2O$$

$M = W$ or Mo

12. CARBENOID REACTIONS[51]

Recently it has been proposed that olefin metathesis, cyclopropanation, *Fischer-Tropsch synthesis,* and other important catalytic reactions proceed through metal-carbene (or metal-carbenoid) intermediates. The olefin metathesis, as shown below, was first found by utilizing some heterogeneous cat-

Carbenoid Reactions

alysts involving oxides of molybdenum, tungsten, or rhenium supported on alumina. Typical types of olefin metathesis are the "triolefin process" and the ring-opening polymerization of cyclopentene to a "polypentenamer." At first the ring-opening polymerization was thought to proceed by the ring opening of the five-membered ring. However, the reaction has now been interpreted to proceed through reactive metal-carbene complexes listed below. The initiation step requires the disproportionation of metal-alkyl complexes.[52]

$$2 \underset{H}{\overset{R}{>}}C=C\underset{H}{\overset{R'}{<}} \underset{\longleftarrow}{\overset{catalyst}{\longrightarrow}} \underset{H}{\overset{R}{>}}C=C\underset{H}{\overset{R}{<}} + \underset{H}{\overset{R'}{>}}C=C\underset{H}{\overset{R'}{<}}$$

1. Triolefin process:

$$2 \; CH_3-CH=CH_2 \longrightarrow CH_3-CH=CH-CH_3 + CH_2=CH_2$$

2. Polymerization of cyclopentene:

$$\text{cyclopentene} \xrightarrow{WCl_6/Et_2AlCl} \text{polypentenamer}_n$$

(a) Initiation:

$$WCl_6 + Et_2AlCl \longrightarrow \{WEtCl_nL_x\} \xrightarrow{disproportionation} \{W(CH-Me)Cl_mL_y\} + C_2H_6$$

(b) Propagation:

$$\left[\underset{}{W\!=\!\!=\!C\overset{R}{\underset{H}{<}}}\right] \longrightarrow \left[W\text{--}CH\text{-}R\right] \longrightarrow \left[W \quad R\right] \longrightarrow \text{polymer chain}_n$$

A number of different catalyst systems have been reported for the olefin metathesis reaction; Table 16 lists selected examples of homogeneous systems. Identity of the metal, as well as auxiliary ligands and reaction conditions strongly influence the catalysis. It is very difficult to draw detailed mechanistic conclu-

Table 16. Representative Examples of Homogeneous Catalysts of Olefin Methathesis at Room Temperature

Catalyst	Ratio	Time	Solvent
WCl_6/EtAlCl$_2$/EtOH	1-4-1	1-3 min	Benzene
WCl_6/n-BuLi/AlCl$_3$	2-4-1	15 min	Benzene
WCl_6/LiAlH$_4$	1-1	15 min	Chlorobenzene
$MoCl_2(NO)_2(PPh_3)_2$/$Me_3Al_2Cl_3$	1-2	2 hr	Chlorobenzene
$ReCl_5$/n-Bu$_4$Sn	2-3	24 hr	Chlorobenzene
$ReCl_5$/Et$_3$Al/O$_2$	1-4	2 hr	Chlorobenzene
Ph$_2$C=W(CO)$_5$	—	8 hr (50°)	Benzene

sions at the present stage. However, the carbene mechanism, which is depicted below, appears to be the most plausible.

$$\left[M=C{\overset{R}{\underset{H}{}}} \right] + {\overset{H}{\underset{R'}{C}}}={\overset{H}{\underset{R'}{C}}}$$

$$\updownarrow$$

$$\left[\begin{array}{c} M\!=\!\!=\!\!C{\overset{R}{}} \\ \vdots \quad \vdots \\ H \\ H \\ \underset{R'}{C}\!=\!\!=\!\underset{R'}{C} \end{array} \right] \rightleftharpoons \left[\begin{array}{c} M \\ \| \\ C \\ H \quad R' \end{array} \right] + RCH\!=\!CHR'$$

active catalyst forming reactions

$$(M\!-\!CH_2\!-\!R \longrightarrow M\!=\!C{\overset{H}{\underset{R}{}}}),$$

$$(M{\overset{C}{\underset{C}{}}} \longrightarrow M\!=\!C{\overset{}{\underset{}{}}}).$$

Deuterated olefins scramble the deuterium label much faster than the olefin metathesis occurs. This indicates that "nonproductive" metathesis is occurring between the two deuterated butene molecules differing only in the isotopic distribution:

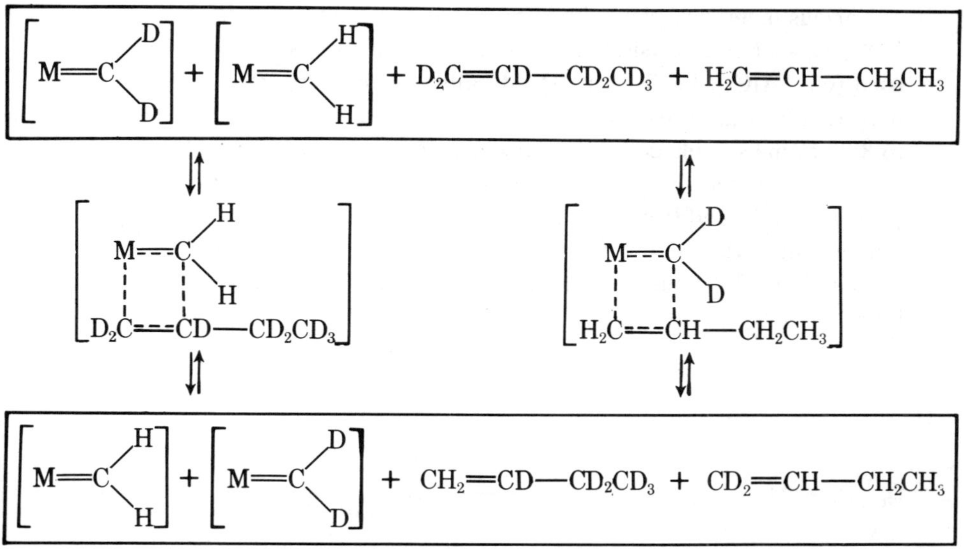

Carbenoid Reactions

$$\begin{bmatrix} M{=}C{\diagup}^{D}_{\diagdown D} \end{bmatrix} + \begin{bmatrix} M{=}C{\diagup}^{H}_{\diagdown H} \end{bmatrix} + D_2C{=}CD{-}CD_2CD_3 + H_2C{=}CH{-}CH_2CH_3$$

$$\updownarrow \qquad\qquad\qquad\qquad \updownarrow$$

$$\begin{bmatrix} M{=\!=\!=}C{\diagdown}^{H}_{H} \\ \vdots \qquad \vdots \\ D_2C{=\!=\!=}CD{-}CD_2CD_3 \end{bmatrix} \qquad \begin{bmatrix} M{=\!=\!=}C{\diagdown}^{D}_{D} \\ \vdots \qquad \vdots \\ H_2C{=\!=\!=}CH{-}CH_2CH_3 \end{bmatrix}$$

$$\updownarrow \qquad\qquad\qquad\qquad \updownarrow$$

$$\begin{bmatrix} M{=}C{\diagup}^{H}_{\diagdown H} \end{bmatrix} + \begin{bmatrix} M{=}C{\diagup}^{D}_{\diagdown D} \end{bmatrix} + CH_2{=}CD{-}CD_2CD_3 + CD_2{=}CH{-}CH_2CH_3$$

Homogeneous metal-complex-catalyzed carbenoid cyclopropanation also proceeds through metal-carbene complexes. The following evidence for this has been obtained: (*a*) there is a high degree of enantioselection in forming chiral cyclopropane derivatives using various optically active metal complexes as catalysts, (*b*) cyclopropanes are formed by the stoichiometric reaction of metal-carbene complexes with some olefins, and (*c*) some catalysts for olefin metathesis also catalyze the formation of cyclopropane derivatives. The following reaction pathways have been proposed[53,54]:

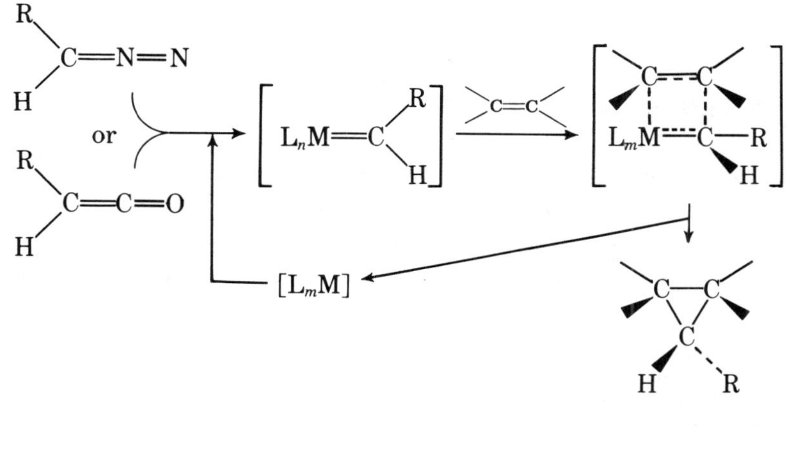

Various metal complexes have been found to be effective for homogeneous cyclopropanation catalysis. Metal halides having high Lewis acidity are generally used (e.g., $AlCl_3$, $FeCl_3$, $ZnCl_2$). Metallic copper, copper(I) oxide, and copper(II) sulfate are the most common and useful *heterogeneous* catalysts at 60 to 80°. Some soluble copper(I) salts [e.g., $Cu(CF_3SO_3)$] are also very active at 25°.[53] Noble metal complex catalysts such as $[(\eta^3\text{-}C_3H_5)PdCl]_2$, $[RhCl(CO)_2]_2$, and $RuCl_2(PPh_3)_3$, are excellent *homogeneous* catalysts at low temperature (10 to 30°). Recently a highly effective enantioselective cyclopropanation of styrene was carried out using optically active cobalt(II) chelates

$$PhCH=CH_2 + N_2CHCO_2R \xrightarrow[-25° \text{ to } 100°]{Co(II) \text{ chelate}} \text{[Ph, } CO_2R\text{ cyclopropane]} + \text{[}CO_2R\text{, Ph cyclopropane]}$$

90% yield

Catalyst:

[Co(III) chelate with $Co^{III}(H_2O)$]$_2$ or [Co(II) chelate with $Co^{II}(H_2O)$]$_2$

13. NITROGEN FIXATIONS[55]

Chemical nitrogen fixation from gaseous nitrogen (N_2, dinitrogen) and hydrogen (H_2, dihydrogen) with the use of homogeneous catalyst systems has been an attractive subject since the discovery of dinitrogen complexes in 1965 and the isolation of pure nitrogenase in 1964.* However no effective homogeneous

---

* The nomenclature using the prefix "di" plus the element is recommended to clearly identify the structure of coordination compounds containing a nitrogen molecule (N_2) or hydrogen molecule (H_2), since there are many compounds involving a nitrogen or a hydrogen atom (N or H) around the coordination sphere.

catalyst that produces ammonia directly from dinitrogen and dihydrogen has been found. Stoichiometric nitrogen fixation was first reported by Vol'pin and Shur in 1964 using a combination of Grignard reagents with transition metal halides of various kinds under nitrogen. Hydrolysis of the reaction mixture gave ammonia in yields of 0.1 to about 70% depending on the identity of the metal species. Titanium, iron, and molybdenum halides have given the best yields.

$$RMgX + MX_n \xrightarrow{N_2} \text{reaction mixture} \xrightarrow[\text{2. alkali}]{\text{1. HCl (aq)}} NH_3$$

Homogeneous systems (e.g., MoO_4^{2-}/Fe^{2+}/thioglycerol/ATP/$NaBH_4$/EtOH) that mimic nitrogenase have been found to reduce isotopically labeled dinitrogen very slowly and very inefficiently to $^{15}NH_3$ under certain conditions. Vanadium(II)/catechol/water systems reduce dinitrogen in a variety of conditions in yields up to about 80% based on the consumed (oxidized) vanadium(II)[56]:

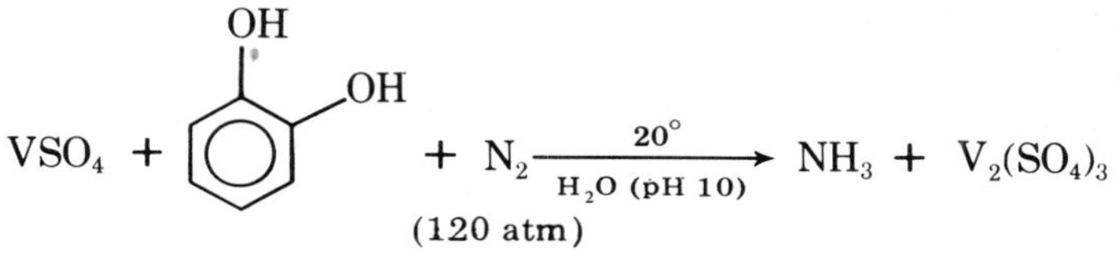

$$VSO_4 + \text{catechol} + N_2 \xrightarrow[\text{H}_2\text{O (pH 10)}]{20°} NH_3 + V_2(SO_4)_3$$
(120 atm)

Although speculative mechanisms for these reactions have been proposed, the absence of conclusive spectral data presently prevents the defining of reaction pathways.

In a further effort to understand or improve the Vol'pin-Shur system $RMgX/MX_n$, various modified systems have been investigated. Of these, Van Tamelen's system $Ti(OR)_4$/Na/naphthalene/THF[57] (Scheme 30) and Vol'pin's $(\eta\text{-}C_5H_5)_2TiX_2$/RMgX/ether system are the best known.[58]

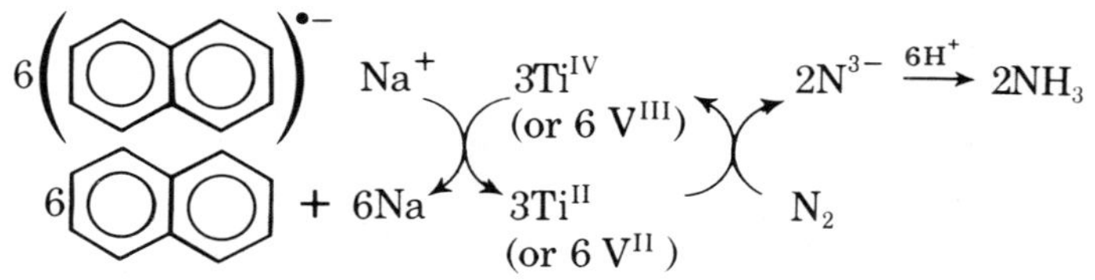

Scheme 30. Van Tamelen's system.

Careful experiments on the interaction of unstable titanocene with gaseous nitrogen in solution have given ir evidence for the formation of an unstable bridged dinitrogen complex $[Cp_2Ti]_2N_2$. Reduction of this unstable species followed by acid hydrolysis has yielded a stoichiometric amount of ammonia. These stoichiometric chemical reductions of dinitrogen have also been performed electrolytically utilizing naphthalene as an electron carrier and aluminum and titanium triisopropoxides as a nitride carrier:

$$N_2 \xrightarrow[\substack{Ti(i\text{-}PrO)_4/NBu_4Cl \\ C_{10}H_8/Al(i\text{-}PrO)_3}]{\text{electrolysis in } CH_3OCH_2CH_2OCH_3} NH_3$$

The use of permethylated η-cyclopentadienyl complexes of titanium and zirconium has been found to give more stable dinitrogen complexes. Low-temperature decomposition of a well-defined zirconium complex

$$(\eta^5\text{-}C_5Me_5)_2\underset{\underset{N_2}{|}}{Zr}-N\equiv N-\underset{\underset{N_2}{|}}{Zr}(\eta^5\text{-}C_5Me_5)_2$$

with HCl has given ammonia (86%) together with some hydrazine.[59] Thus the mechanism of the Vol'pin-Shur reaction appears to involve the formation of an unstable dinitrogen complex in which the dinitrogen is activated for protolytic reaction.

Since the first preparation of $[Ru(NH_3)_5N_2]^{2+}$, a dinitrogen complex of Ru(II), from hydrazine hydrate and $RuCl_3$, a variety of dinitrogen complexes have been prepared (see Table 17). The first reversibly bound dinitrogen complex $CoH(N_2)(PPh_3)_3$ was observed in the reaction of $Co(acac)_3/PPh_3/AlEt_2(OEt)$, $CoCl_2/NaBH_4/PPh_3/EtOH$, or $H_3Co(PPh_3)_3$ with dinitrogen.

Dinitrogen itself is difficult to reduce, and even when coordinated, it resists chemical reduction. Protonation of molybdenum (0) or tungsten (0) dinitrogen complexes have been found after several steps to produce ammonia as a final product. The yield of ammonia has been as high as 85% for the reaction of $W(N_2)_2(PMe_2Ph)_4$ with H_2SO_4 in MeOH.[60]

$$cis\text{-}W(N_2)_2(PMe_2Ph)_4 + H_2SO_4 \xrightarrow[\text{MeOH}]{\text{excess}} NH_3, 85\%$$

The following stoichiometric protonations with hydrogen halides have given stable diazenyl ($=N-NH_2$) complexes (Fig. 54). Although it seems to be an intermediate stage for the production of N_2H_4 or ammonia, neither further

Table 17. Typical Examples of Dinitrogen Complexes[55]

End-on type (M—N≡N)
  Matrix-isolated species (thermally very unstable)
    $M(N_2)_n$: $n = 1-4$; M = Ni, Pd, Pt
    $Ni(CO)_n(N_2)_m$, $M(N_2)_2(O_2)$: = M = Ni, Pd, Pt
  Stabilized by phosphines
    $CoH(N_2)(PPh_3)_3$, $RhH(N_2)(P(t-Bu)_2Ph)_2$, $IrCl(N_2)(PPh_3)_2$
    $FeH_2(N_2)(PPh_3)_3$, $RuH_2(N_2)(PPh_3)_3$
    $[ReCl(N_2)(PMe_2Ph)_4]^{0,+1}$
    trans-$Mo(N_2)_2(diphos)_2$, cis-$W(N_2)_2(PMe_2Ph)_4$
  Stabilized by η^5-C_5H_5
    $(C_5Me_5)_2Ti(N_2)$, $(C_5H_5)_2ZrN_2$
    $(C_5H_5)Mn(N_2)(CO)_2$
  Stabilized by NH_3
    $[Ru(N_2)(NH_3)_5]^{2+}$, $[Os(N_2)(NH_3)_5]^{2+}$, cis-$[Os(N_2)_2(NH_3)_4]^{2+}$
Bridging type (M—N≡N—M')
  $\{[(C_6H_{11})_3P]_2Ni\}_2N_2$, $(Me_2PhP)_4ClRe(N_2)MoCl_4(THF)$
  $((C_5H_5)_2Ti)_2N_2$, $[(C_5Me_5)_2Zr]_2N_2$, $[(NH_3)_5RuN_2Os(NH_3)_5]^{2+}$

Side-on type (M···|||$\genfrac{}{}{0pt}{}{N}{N}$)
  $[Ni_2(N_2)(PhLi)_6(Et_2O)_2]_2$

protonation nor reduction of the diazenyl complexes has yielded ammonia:

$$W(N_2)_2(Ph_2PCH_2CH_2PPh_2)_2 \xrightarrow[THF]{2HCl} WH(N_2)_2(dpe)_2$$
$$(dpe) \qquad \downarrow HCl$$
$$\xrightarrow[THF]{>6HX} WX_2(=N-NH_2)(dpe)_2$$
$$X = Cl, Br$$

$$Mo(N_2)_2(dpe)_2 \xrightarrow[\text{in excess}]{HCl} MoH_2Cl_2(dpe)_2 + 2N_2$$
$$\xrightarrow{6HBr} MoBr_2(=N-NH_2)(dpe)_2$$

The use of oxyacids rather than halogen acids in protonations is essential for the formation of ammonia from coordinated dinitrogen. All the above-mentioned nitrogen fixation reactions involve transition metals coordinated with *tert*-phosphines, η^5-cyclopentadienyl groups, or other normal anionic inorganic li-

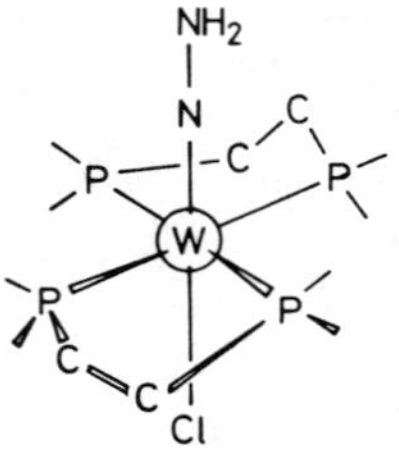

Fig. 54. The x-ray structure of $[W(N_2H_2)Cl(Ph_2PCH_2CH_2PPh_2)_2]^+[BPh_4]^-$.

gands. On the other hand, investigation of the active site(s) of the nitrogenase enzyme has revealed one or two molybdenum (III–V) ions in the vicinity of a ferredoxin-type iron-sulfur prosthetic group. The active molybdenum has been thought to be coordinated with *sulfur* ligands. One proposed mechanism has invoked a hypothetical side-on (or η^2-) dinitrogen coordination to the reduced molybdenum, but others suggest that cooperative action of both iron and molybdenum activates dinitrogen and leads to the reductive cleavage occurring at the enzyme active site. These possibilities can be represented as follows:

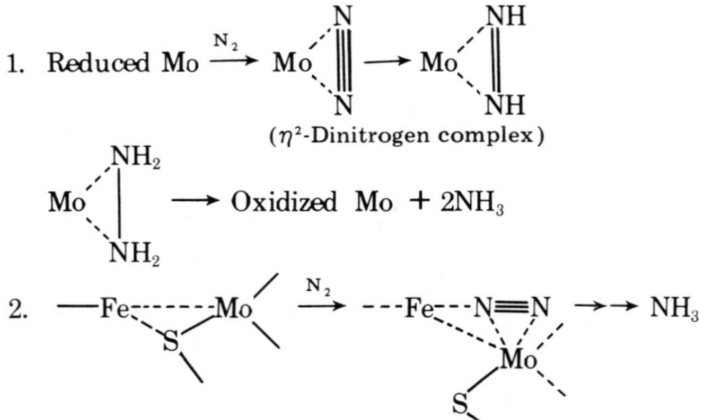

1. Reduced Mo $\xrightarrow{N_2}$ Mo(N≡N) $\longrightarrow$ Mo(NH=NH)

                                         (η^2-Dinitrogen complex)

    Mo(NH$_2$)(NH$_2$) $\longrightarrow$ Oxidized Mo + 2NH$_3$

2. —Fe–S–Mo $\xrightarrow{N_2}$ —Fe⋯N≡N⋯Mo–S $\longrightarrow$ NH$_3$

The role of the sulfur ligands in the enzymic nitrogen fixation is not well understood, and chemistry of these sulfur chelates or clusters of iron and molybdenum is being studied further. Preliminary results on catalysis of some sulfur-chelated binuclear molybdenum complexes in the reductive cleavage of the N=N bond in azobenzene should encourage research in this field.[62a]

$$\text{PhN}=\text{NPh} + \text{NaBH}_4 \xrightarrow[\text{Mo-S chelates}]{25°\ \text{EtOH}} \text{PhNH}_2$$

For example,

$$R-C\begin{smallmatrix}S\\S\end{smallmatrix}\overset{O}{\underset{S}{\overset{\|}{Mo}}}\begin{smallmatrix}S\\S\end{smallmatrix}\overset{O}{\underset{S}{\overset{\|}{Mo}}}\begin{smallmatrix}S\\S\end{smallmatrix}C-R$$

$R = NR'_2, OR'$

Very recently x-ray absorption edge measurements and extended x-ray absorption fine structure (EXAFS) measurements have given important information on metal-ligand bond lengths and identification of the coordinating atoms. The molybdenum active site of nitrogenase has thus been found to be a mixed molybdenum-iron-sulfur cluster with Mo—S bond lengths of 2.35 and 2.49 Å, and a Mo—Fe bond length of 2.72 Å.[62b]

14. ENANTIOSELECTIVE CATALYTIC REACTIONS[63,64]

One of the most important and interesting illustrations of the high selectivity of homogeneous catalysis is the enantioselective catalytic reaction by chiral complexes. When one considers stoichiometry of chiral products, the importance of effective chiral catalysts lies in their ability to produce hundred- or thousandfold amounts of chiral products starting from achiral reactants. Some early research examined the enantioselective efficiency by means of metal complexes involving optically active amines, alcohols, Schiff bases, or amino acids, and resulted in relatively low optical yields (generally 1–6%) of chiral products.

Efficient chiral catalysts were found only after the pioneering work by Horner and Mislow, who prepared optically active chiral *tert*-phosphines. These were combined with the discovery of the excellent catalytic properties of Wilkinson's complex $RhCl(PPh_3)_3$. Knowles and others have reported exceedingly good optical yields of up to 95% from the catalytic hydrogenation of very sterically hindered and multifunctional olefins such as

$$\text{PhCH}=\underset{\underset{\text{NHCOCH}_3}{|}}{\text{C}}-\text{CO}_2\text{H}.$$

Some chiral phosphines, such as one having an *o*-anisyl group on the chiral phosphorus atom, have been found to be exceptionally effective as chiral ligands. Also, a novel optically pure, chelating diphosphine was synthesized from readily available (+) or (−)-tartaric acid by Kagan and Dang.[65a] When combined with Rh(I), this diphosphine has been found to be very effective not only in olefin hydrogenation but also the hydrosilylation of olefins or ketones.[65]

Asymmetric ferrocenylphosphines also have been superb chiral ligands for the same reaction (maximum optical yield, 97%).

Enantioselective Catalytic Reactions

The center of chirality does not necessarily have to be a near neighbor of the active metal center to produce good optical yields. Easily available *tert*-phosphines, such as diphenyl(*l*-neomenthyl)phosphine, have been found to give a highly enantioselective yield in the hydrogenation of (*E*)-β-methylcinnamic acid.[66]

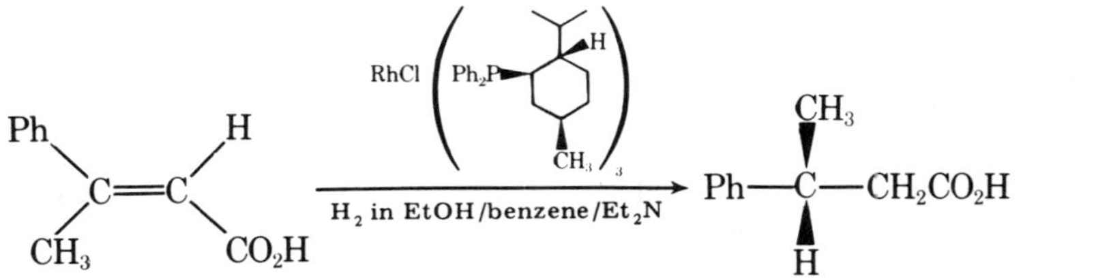

61% ee (enantiomeric excess)

In view of the tedious steps (see Scheme 31) that have been needed for the preparation of *P-chiral tert*-phosphines, the use of *side-chain chiral* phosphines has a distinct advantage. However such catalysts have not shown high selectively in the hydrogenation of substrates other than (*E*)-β-methylcinnamic acid. A slight change of substrate structure seems to exert great influence over the enantioselectivity in an unknown way. Typical preparative reactions of chiral phosphines are shown in Schemes 31 and 32.

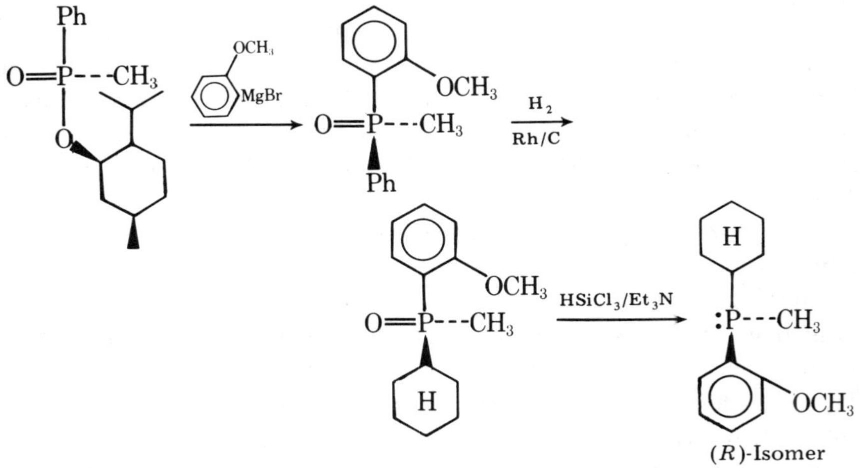

Scheme 31. The preparation of (*R*)-P(CH$_3$)(cyclohex)(o-anisyl).

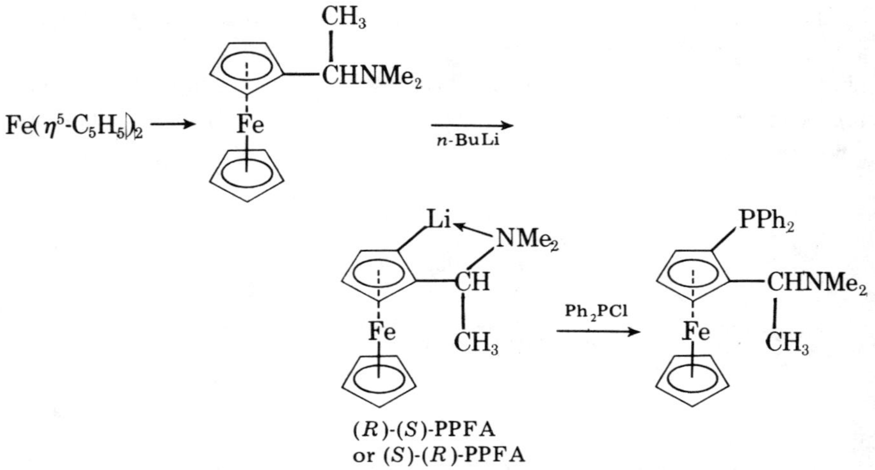

$Fe(\eta^5-C_5H_5)_2 \longrightarrow$ [Ferrocene with CHNMe₂ and CH₃ substituent] $\xrightarrow{n\text{-BuLi}}$ [Lithiated ferrocene intermediate] $\xrightarrow{Ph_2PCl}$ [PPh₂-substituted ferrocene]

(R)-(S)-PPFA
or (S)-(R)-PPFA

Scheme 32. The preparation of optically active ferrocenyl phosphines.[67]

Recently further new effective chiral diphosphines (*P—P) have been developed, and the new chiral catalysts (Rh¹/*P—P) have been utilized to prepare important chemicals such as amino acid derivatives and panthothenic acid.[68-73]

As described above, chelating chiral phosphines are important as ligands in cationic Rh(I) complexes that catalyze highly enantioselective olefin hydrogenation. The catalyst precursor [Rh(cod)(P—P*)]⁺ (cod, 1,5-cyclooctadiene; P—P*; chiral diphosphine), is first activated by reaction with 2 moles of H_2 to labile [Rh(P—P*)(solv)₂]⁺ (solv = solvent molecule, e.g., MeOH or acetone), which then reacts with the functional olefin (f-Ol), for example, acetamidocinnamic acid (PhCH=C(NHCOCH₃)CO₂H), to give [Rh(P—P*)(f-Ol)]⁺.[74] The structure of this important catalytic intermediate has recently been investigated by x-ray analysis and ^{31}P nmr spectroscopy and has indicated an interesting correlation of the enantioselective trend with the conformation of the chelate ring. It is known that the chirality at the chelate ring induces the conformation, δ or λ; therefore each two phenyl groups on the P atoms are forced to take "face" or "edge" positions as shown in Scheme 33.

These particular positions of the four phenyl groups seem to determine the enantioselective trend in catalytic hydrogenation of acetamidocinnamic acid by preferential coordination of one enantioface of this prochiral olefin as Table 18 indicates.

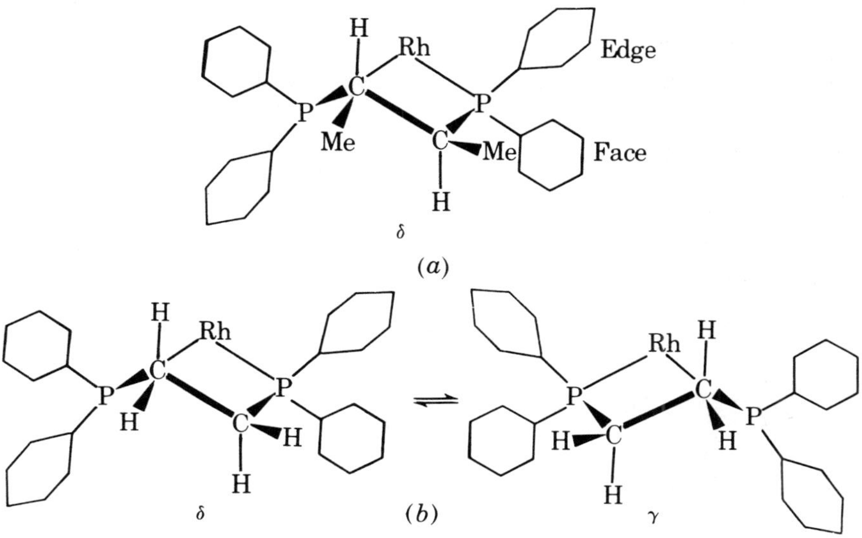

Scheme 33. (a) (S,S)-Chiraphos complex.[69] (b) Ph$_2$PCH$_2$CH$_2$PPh$_2$ Complex.

Enantioselective hydroformylation has been examined using a combination of RhH(CO)$_2$(PPh$_3$)$_2$ with some known optically active *tert*-phosphines. The optical yield of the product has been consistently lower than those observed in the hydrogenation using the same chiral phosphine. This low selectivity of the hydroformylation may have been a result of the well-known multistep nature of the reaction.[75]

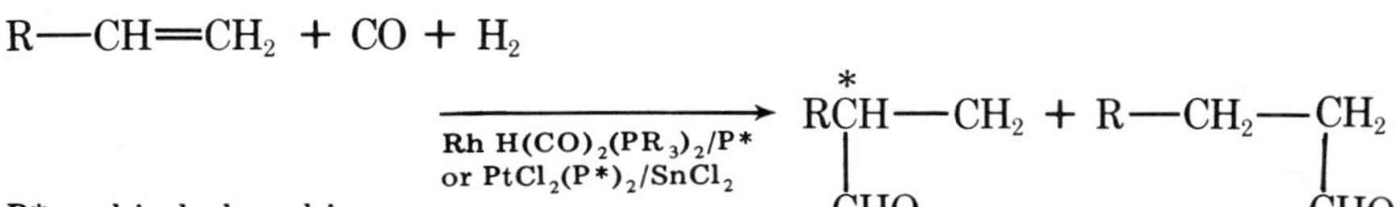

$$R-CH=CH_2 + CO + H_2 \xrightarrow[\text{or PtCl}_2(P^*)_2/\text{SnCl}_2]{\text{Rh H(CO)}_2(\text{PR}_3)_2/P^*} R\overset{*}{C}H-CH_2 + R-CH_2-CH_2$$
$$\underset{\text{CHO}}{|} \qquad \underset{\text{CHO}}{|}$$

P* = chiral phosphine

Using a novel chiral diphosphine (see below), the hydroformylation of styrene has yielded a branched aldehyde in 45% optical yield.[76]

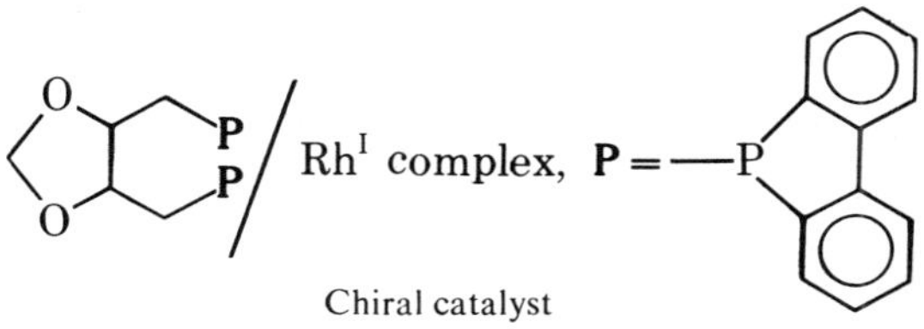

Rh^I complex, **P** =

Chiral catalyst

Table 18. Optical Yields with Selected Chiral Chelate Catalysts

Chiral Chelate	Configuration of Excess Enantiomer	Optical Yield (%)
(R,R)-DIPAMP	S	94
(S,S)-Chiraphos	R	89
(R)-Prophos	S	90
(S,S)-DIOP	R	82
(S)-(R)-BPPFA	S	93
(S,S)-BPPM	R	91

Enantioselective Catalytic Reactions

Alkylations of some olefins with ethylene have also been performed enantioselectively (see Scheme) by using the highly active catalyst system [NiCl(C_3H_5)]$_2$/Al$_2$Cl$_3$Et$_2$/2P* at -30 to $-78°$. A specially designed bulky chiral *tert*-phosphine, P(CH_3)(*i*-Pr)(*l*-menthyl), has yielded the best result (80% ee at $-78°$).[77]

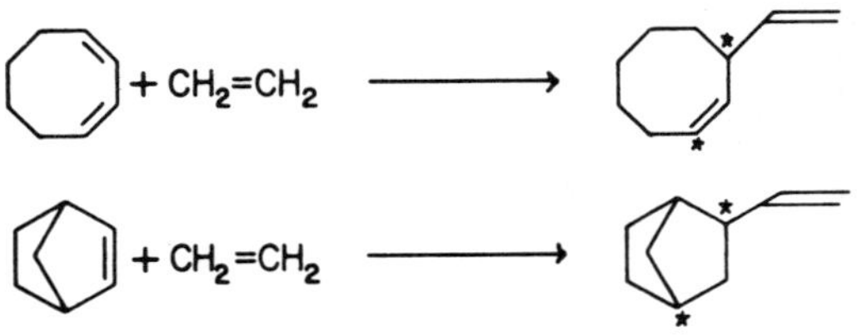

Scheme 34

An x-ray structural analysis of the enantiomeric nickel complex shown below has revealed that the one face of the coordination plane is effectively blocked from attack.

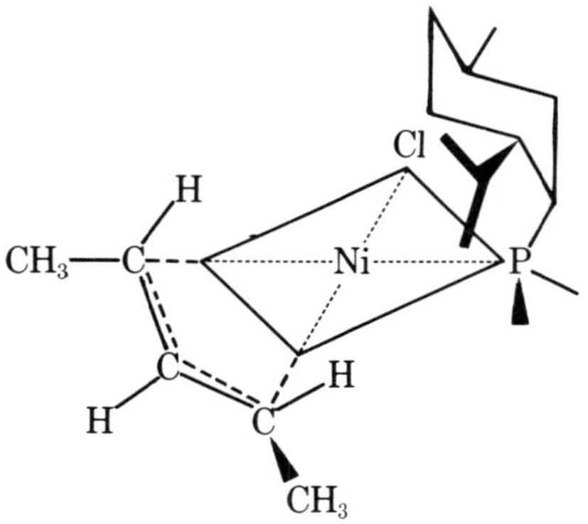

Although enantioselective catalytic cyclopropanation has been tried using copper complexes with optically active ligands, the optical yields have generally been low (3–6%). However remarkable improvement of the optical selectivity has been made recently by use of chiral cobalt(II) chelates that have been prepared from *d*-camphor,[54] and by use of specially designed tridentate chiral chelates for Cu(II).[78] The former catalyst has been readily prepared from natural *d*-camphor and has caused an enantioselective carbenoid cyclopropanation of

Table 19. Asymmetric Carbenoid Cyclopropanation of Various Olefins with Ethyl Diazoacetate

Olefin Substrate	Catalyst (mol %)	Temperature (°C)	Yield (%)	Product	Configuration	Optical Yield (%)
PhCH=CH$_2$	2.8	0	92	Ph, CO$_2$Et (cyclopropane)	1S, 2R	~70
				Ph, CO$_2$Et (cyclopropane)	1S, 2S	75
Ph$_2$C=CH$_2$	2.5	0	95	Ph, Ph, CO$_2$Et (cyclopropane)	(1S)	70
Ph(MeO$_2$C)C=CH$_2$	2.1	0	92	Ph, CO$_2$Me, CO$_2$Et (cyclopropane)	(1R, 2S)	37
				Ph, CO$_2$Et, CO$_2$Me (cyclopropane)	(1S, 2S)	71

Data of Nakamura et al., in ref. 54.

styrene in 75% optical yield. Selected results appear in Table 19. The latter catalyst has been effective for the enantioselective synthesis of chrysanthemic acid in an optical yield amounting to 90% (see p. 25 and 192).

$$\text{(dimethylallyl alkene)} + N_2CHCO_2R \xrightarrow{\text{chiral Cu chelate}} \text{(cyclopropane product)}$$

R = *l*-menthyl   Maximum optical yield, 90%

An effective enantioface selection of the intermediate metal-carbene species

$$M=C\begin{array}{c}R\\ \\H\end{array},$$

is thus achieved by the chiral chelate ligands surrounding the metal.

SELECTED READINGS

A) Enzymes

H. R. Mahler and E. H. Cordes, *Biological Chemistry,* Harper & Row, New York, 1966.

A. Mazur and B. Harrow, *Textbook of Biochemistry,* W. B. Saunders Co., Philadelphia, 1971.

M. L. Bender and L. J. Brubacher, *Catalysis and Enzyme Action,* McGraw-Hill, New York, 1973.

M. N. Hughes, *The Inorganic Chemistry of Biological Processes,* Wiley, New York, 1970.

G. L. Eichhorn, Ed., *Inorganic Biochemistry,* Vols. 1 and 2, Elsevier, Amsterdam, 1973.

B. S. Cooperman, in *Metal Ions in Biological Systems,* H. Sigel, Ed., Vol 5, Marcel Dekker, New York, 1976.

J. M. Pratt, *Inorganic Chemistry of Vitamin B_{12},* Academic Press, London 1972.

B) Organometallic Catalysts

B. R. James, *Homogeneous Hydrogenation,* Wiley-Interscience, New York, 1972.

S. Otsuka and A. Nakamura, "Acetylene and Allene Complexes; Their Implication in Homogeneous Catalysis," *Adv. Organomet. Chem.,* **14,** 245 (1976).

G. Schrauzer, Ed. *Transition Metals in Homogeneous Catalysis,* Academic, New York, 1971.

C. W. Bird, *Transiton Metal Intermediates in Organic Synthesis,* Logos Press, London, 1967.

A. E. Vol'pin and B. Shur, *Organometallic Reactions,* Vol. 1, Academic, New York, 1972, p. 1.

The Chemistry and Biochemistry of Nitrogen Fixation, J. R. Postgate, Ed., Plenum Press, London, 1971, Chapter 2; G. L. Leigh, "A Biological N_2 Fixation," p. 19, Chapter 3; J. Chatt and R. L. Richards, "N_2 Complexes & N_2 Fixation," p. 57–101.

J. P. Kennedy, *Cationic Polymerization of Olefins,* Wiley, New York, 1975.

I. Wender and P. Pino, Eds., *Organic Syntheses via Metal Carbonyls,* Vol. 1, 1968, and Vol. 2, 1977, Wiley-Interscience, New York.

REFERENCES

1. M. L. Bender, *Mechanisms of Homogeneous Catalysis from Protons to Proteins*, Wiley, New York, 1971.
2. W. P. Jencks, *Catalysis in Chemistry and Enzymology*, McGraw-Hill, New York, 1969.
3. L. Senatore, E. Ciuffarin, M. Isola, and M. Vichi, *J. Am. Chem. Soc., 98,* 5306 (1976).
4. D. R. Storm and D. E. Koshland, Jr., *J. Am. Chem. Soc., 94,* 5815 (1972). See also, R. M. Moriarty and T. Adams, *J. Am. Chem. Soc., 95,* 4070 (1973).
5. R. E. Dickerson and I. Geis, *The Structure and Action of Proteins*, Benjamin, Menlo Park, Calif., 1969.
6. F. Basolo and R. G. Pearson, *Mechanisms of Inorganic Reactions*, 2nd ed., Wiley, New York, 1967, p. 217.
7. Ref. 2, p. 112.
8. S. H. Goh, L. E. Closs, and G. L. Closs, *J. Org. Chem., 34,* 25 (1969).
9. W. L. Jörgensen and L. Salem, *The Organic Chemist's Book of Orbitals*, Academic Press, New York, 1973, p. 126.
10. J. E. Coleman, in *Inorganic Biochemistry*, Vol. 1, G. L. Eichhorn, Ed., Elsevier, Amsterdam, 1973, p. 488.
11. H. B. Bürgi, J. M. Lehn, and G. Wipff, *J. Am. Chem. Soc., 96,* 1956 (1974).
12. G. Henrici-Olivé and S. Olivé, *Angew. Chem., Int. Ed., 2,* 873 (1967).
13. G. N. Schrauzer, *Angew. Chem., Int. Ed., 14,* 514 (1975).
14. A. White, P. Handler, and E. L. Smith, *Principles of Biochemistry*, 5th ed., McGraw-Hill, New York, 1973, p. 333.
15. R. W. Dobson and B. Warnqvist, *Inorg. Chem., 10,* 2624 (1971).
16. R. Cramer, *Acc. Chem. Res., 1,* 186 (1968).
17. H. C. Clark and H. Kurosawa, *Inorg. Chem., 11,* 1275 (1973).
18. R. G. Miller, P. A. Pinke, R. D. Stauffer, H. J. Golden, and D. J. Baker, *J. Am. Chem. Soc., 96,* 4211 (1974).
19. P. A. Pinke and R. G. Miller, *J. Am. Chem. Soc., 96,* 4221 (1974).
20. See recent reviews: K. C. Bishop, III, *Chem. Rev., 76,* 461 (1976); K. Fukui and S. Inagaki, *J. Am. Chem. Soc., 97,* 4445 (1975); R. Noyori, M. Yamakawa, and H. Takaya, *J. Am. Chem. Soc., 98,* 1471 (1976).
21. (a) For details see B. R. James, *Homogeneous Hydrogenation*, Wiley, New York, 1973. (b) C. A. Tolman, P. Z. Meakin, D. L. Lindner, and J. P. Jesson, *J. Am. Chem. Soc., 96,* 2762 (1974).
22. T. Yoshida and S. Otsuka, *J. Am. Chem. Soc.,* to be published; K. Achiwa, T. Kogure, and I. Ojima, *Chem. Lett., 1979,* 297.
23. E. L. Muetterties, F. J. Hirsekorn, and M. C. Rakowski, *J. Am. Chem. Soc., 97,* 237 (1975); *100,* 2405 (1975); P. M. Maitlis, *Acc. Chem. Res., 11,* 301 (1978).
24. M. Siskin and J. Poncelli, *J. Am. Chem. Soc., 96,* 3640 (1974); J. Wristers, *J. Am. Chem. Soc., 97,* 4312 (1975).

References

25. H. M. Feder and J. Halpern, *J. Am. Chem. Soc.*, 97, 7186 (1975).
26. G. Platbrood and L. Wilputte-Steinert, *J. Mol. Catal.*, 1, 263 (1976); M. F. Farona, *Organomet. React. Synth.*, 6, 246 (1977).
27. J. Hjortkjaer and V. W. Jensen, *Ind. Eng. Chem. PRD*, 15, 46 (1976); P. Pino, G. Braca, G. Sbrana, and A. Cuccuru, *Chem. Ind. (Milan)*, 1732 (1968).
28. (a) C. W. Bird, *Transition Metal Intermediates in Organic Synthesis*, Logos Press, London, 1967, p. 117. (b) M. van Boven, N. H. Alemdaroglu, and J. M. L. Penninger, *IEC Prod. Res. Dev.*, 14, 259 (1975).
29. (a) J. Tsuji, *Adv. Org. Chem.*, 6, 109 (1969). (b) D. Forster, *J. Am. Chem. Soc.*, 98, 846 (1976).
30. J. D. McClure and K. W. Barnett, *J. Organomet. Chem.*, 80, 385 (1974).
31. P. W. Jolly and G. Wilke, *The Organic Chemistry of Nickel*, Vols. 1 and 2, Academic Press, New York, 1974.
32. P. Heimbach, P. W. Jolly, and G. Wilke, *Adv. Organomet. Chem.*, 8, 29 (1970).
33. P. Heimbach, *Angew. Chem., Int. Ed.*, 12, 975 (1973).
34. A. Nakamura and S. Otsuka, *Adv. Organomet. Chem.*, 14, 245 (1976); *J. Am. Chem. Soc.*, 94, 1037 (1972).
35. (a) Y. Tohda, K. Sonogashira, and N. Hagihara, *J. Organomet. Chem.*, 110, C53 (1976). (b) G. Wilke, *Pure Appl. Chem.*, 50, 677 (1978).
36. H. Yamazaki and Y. Wakatsuki, *Tetrahedron Lett.*, 3383 (1973); *J. Am. Chem. Soc.*, 95, 5781 (1973).
37. Review: N. M. Bikales, in *Homogeneous Catalysis*, B. J. Luberoff, Ed., American Chemical Society, *Advances in Chemistry* Series, Vol. 70, ACS, Washington, D.C., 1968, p. 233. See also S. M. Atlas and H. F. Mark, *Catal. Rev.*, 13, 1 (1976); *Coordination Polymerization*, J. C. W. Chien, Ed., Academic Press, New York, 1975.
38. G. Henrici-Olivé and S. Olivé, *Angew. Chem., Int. Ed.*, 9, 243 (1970); 10, 105 (1971).
39. D. G. H. Ballard, *Adv. Catal.*, 23, 263 (1973); in *Coordination Polymerization*, J. C. W. Chien, Ed., Academic Press, New York, 1975, p. 223.
40. (a) Y. Doi, J. Kinoshita, A. Morinaga, and T. Keii, *J. Polym. Sci., A*, 13, 2491 (1975). (b) K. J. Ivin, J. J. Rooney, C. D. Stewart, M. L. H. Green, and R. Mahtab, *Chem. Commun.*, 604 (1978).
41. (a) J. P. Durand and Ph. Teyssié, *J. Polymer Sci., B6*, 299 (1968). (b) J. Furukawa, K. Haga, Y. Ishida, T. Yoshimoto, and K. Sakamoto, *Polym. J.*, 2, 371 (1971). (c) *Chem. Eng. News*, April 24, 1978. (d) C. K. Chiang, M. A. Druy, S. C. Gau, A. J. Heeger, E. J. Louis, A. G. MacDiarmid, Y. W. Park, and H. Shirakawa, *J. Am. Chem. Soc.*, 100, 1013 (1978). (e) H. Shirakawa and S. Ikeda, *Polym. J.*, 2, 231 (1971). (f) H. Shirakawa, T. Ito, and S. Ikeda, *Polym. J.*, 4, 460 (1973). (g) T. Ito, H. Shirakawa, and S. Ikeda, *J. Polym. Sci., Polym. Chem.*, 12, 11 (1974). (h) S. Otsuka and A. Nakamura, and K. Tani, *Kogyo Kogaku Zasshi*, 70, 2007 (1967). (j) S. Otsuka, A. Nakamura, and K. Tani, *Kogyo Kagaku Zasshi*, 72, 1809 (1969); F. L. Bowden and R. Giles, *Coord. Chem. Rev.*, 20, 81 (1976).
42. J. E. Lyons, in *Homogeneous Catalysis—II*, D. Forster and J. F. Roth, Eds., American Chemical Society *Advances in Chemistry* Series, Vol. 132, ACS, Washington, D.C., 1974, p. 64; P. M. Henry and R. N. Pandey, *ibid.*, p. 33; R. G. Brown and J. M. Davidson, ibid., p. 49.

43. J. E. Lyons, *Aspects of Homogeneous Catalysis*, R. Ugo, Ed., Vol. 3, Reidel, Dordrecht, Holland, 1977, p. 3; J. K. Kochi, *Organometallic Mechanisms and Catalysis*, Academic Press, New York, 1978, p. 69.
44. Y. Fujiwara, I. Moritani, S. Danno, R. Asano, and S. Teranishi, *J. Am. Chem. Soc., 91*, 7166 (1969).
45. R. A. Sheldon and J. K. Kochi, *Adv. Catal.*, 25, 272 (1976); D. E. Webster, *Adv. Organomet. Chem.*, 15, 147 (1977).
46. J. Tsuji and H. Takayanagi, *J. Am. Chem. Soc., 96*, 7349 (1974), *Tetrahedron Lett.*, 1975, 1245; M. M. Rogic, T. R. Demmin, and W. B. Hammond, *J. Am. Chem. Soc., 100*, 5472 (1978), *Biological Hydrozylation Mechanisms*, G. S. Boyd and R. M. S. Smillie, Eds., Academic Press, New York, 1972.
47. J. E. Lyons, in *Fundamental Research in Homogeneous Catalysis*, M. Tsutsui and R. Ugo, Eds., Plenum Press, New York, 1977.
48. R. Osterberg, *Coord. Chem. Rev., 13*, 309 (1974); A. Nishinaga, T. Tojo, and T. Matsu-ura, *Chem. Commun.*, 896 (1974); A. Avdeef and W. P. Schafer, *J. Am. Chem. Soc., 98*, 5153 (1976). Reviews: A. J. Fee, *Structure and Bonding*, Vol. 23, Springer, Heidelberg, 1976, p. 1.; G. A. Hamilton, P. K. Adolf, J. de Jersey, G. C. DuBois, G. R. Dyrkacz, and R. D. Libby, *J. Am. Chem. Soc., 100*, 1899 (1978).
49. H. Kojima, S. Takahashi, and N. Hagihara, *Chem. Commun.*, 230 (1973).
50. (a) S. Otsuka, A. Nakamura, and Y. Tatsuno, *Chem. Commun.*, 836 (1967). (b) A. Sen and J. Halpern, *J. Am. Chem. Soc., 99*, 8337 (1977).
51. L. J. Haines and G. J. Leigh, *Chem. Soc. Rev., 4*, 155 (1975); J. C. Mol and J. A. Moulyjn, *Adv. Catal., 24*, 131 (1975). W. B. Hughes, *Organomet. Chem. Synth. 1*, 341 (1972); W. B. Hughes, in *Homogeneous Catalysis—II*, D. Forster and J. F. Roth, Eds., American Chemical Society *Advances in Chemistry* Series, Vol. 132, ACS, Washington, D.C., 1974, p. 192; N. Calderon, E. A. Ofstead, and W. A. Judy, *Angew. Chem., 88*, 433 (1976); T. J. Katz, *Advan. Organomet. Chem., 16*, 283 (1977).
52. R. R. Schrock and G. W. Parshall, *Chem. Rev., 76*, 243 (1976); P. J. Davidson, M. F. Lappert, and R. Pearce, *Chem. Rev., 76*, 219 (1976).
53. R. G. Salomon and J. K. Kochi, *J. Am. Chem. Soc., 95*, 3300 (1973).
54. A. Nakamura, A. Konishi, Y. Tatsuno, and S. Otsuka, *J. Am. Chem. Soc., 100*, 3443 (1978); A. Nakamura, A. Konishi, R. Tsujitani, and S. Otsuka, *J. Am. Chem. Soc., 100*, 3449 (1978).
55. A. D. Allen, R. O. Harris, B. R. Loescher, J. R. Stevens, and R. N. Whiteley, *Chem. Rev., 73*, 11 (1973); J. Chatt and G. J. Leigh, *Chem. Soc. Rev., 1*, 121 (1972); J. Chatt, J. R. Dilworth, and R. L. Richards, *Chem. Rev., 78*, 589 (1978); P. Giannoccaro, M. Rossi, and A. Sacco, *Coord. Chem. Rev., 8*, 77 (1972); E. L. Moorehead, P. R. Robinson, T. M. Vickrey, and G. N. Schrauzer, *J. Am. Chem. Soc., 98*, 6555 (1976); *Proceedings of the 1st International Symposium on Nitrogen Fixation*, Vol. 1, W. E. Newton and C. J. Nyman, Eds., Washington, State University Press, Pullman, 1976; A. E. Shilov, *Usp. Khim., 43*, 863 (1974); *The Chemistry and Biochemistry of Nitrogen Fixation*, J. R. Postgate, Ed., Plenum Press, London and New York, 1971, Chapter 2; G. J. Leigh, "A Biological N_2 Fixation," *ibid.*, p. 19, J. Chatt and R. L. Richards, "N_2 Complexes and N_2 Fixation," *ibid.*, pp. 57–101.

References

56. L. A. Nikonova, A. G. Ovcharenko, O. N. Efimov, V. A. Avilov, and A. E. Shilov, *Kinet. Katal.*, *13*, 1602 (1972); L. A. Nikonova, N. I. Pershikova, M. V. Bodeiko, L. G. Olijnik, D. N. Sokolov, and A. E. Shilov, *Dokl. Acad. Nauk SSSR*, *216*, 1 (1974).
57. E. E. van Tamelen, R. B. Fechter, S. W. Schneller, G. Boche, R. H. Greeley, and B. Åckermark, *J. Am. Chem. Soc.*, *91*, 1551 (1969).
58. M. E. Vol'pin and V. B. Shur, *Organomet. React.*, *1*, 55 (1970).
59. (a) J. M. Manriquez, R. D. Sanner, R. E. Marsh, and J. E. Bercaw, *J. Am. Chem. Soc.*, *98*, 3042 (1976). (b) R. D. Sanner, J. M. Manriquez, R. E. March, and J. E. Bercaw, *J. Am. Chem. Soc.*, *98*, 8351 (1976).
60. J. Chatt, A. J. Pearman, and R. L. Richards, *Nature*, *259*, 204 (1976); *235*, 39 (1975); *J. Organomet. Chem.*, *101*, C45 (1976); *J. Chem. Soc. Dalton, Trans.*, 1852 (1977).
61. G. A. Heath, R. Mason, and K. M. Thomas, *J. Am. Chem. Soc.*, *96*, 259 (1974).
62. (a) A. Nakamura, K. Sugihashi, and S. Otsuka, *J. Less-Common Met.*, *54*, 495 (1977). (b) S. P. Cramer, K. O. Hodgson, W. D. Gillum, and L. E. Mortenson, *J. Am. Chem. Soc.*, *100*, 3398 (1978).
63. O. Korpium, R. A. Lewis, J. Chickos, and K. Mislow, *J. Am. Chem. Soc.*, *90*, 4842 (1968); T.-P. Dang and H. B. Kagan, *J. Am. Chem. Soc.*, *94*, 6429 (1972); W. S. Knowles, M. J. Sabacky, and B. D. Vineyard, *Chem. Commun.*, 1445 (1969); 10 (1972); *Chem. Tech.*, *1972*, 590; W. S. Knowles, M. J. Sabacky, and B. D. Vineyard, *Homogeneous Catalysis—II*, D. Forster and J. F. Roth, Eds., American Chemical Society *Advances in Chemistry* Series, Vol. 132, ACS, Washington, D.C., 1974, p. 274.
64. Reviews: J. D. Morrison, W. F. Master, and M. K. Neuberg, *Adv. Catal.*, *25*, 81 (1976); J. W. Scott and D. Valentine, Jr., *Science*, *184*, 943 (1974); I. Ojima, K. Yamamoto, and M. Kumada, *Aspects of Homogeneous Catalysis*, R. Ugo, Ed., Vol. 3, Reidel, Dordrecht, Holland, p. 186; J. Solodar, *Chem. Tech.*, 590 (1972); K. Yamamoto, *Kagaku to Kogyo*, *26*, 193 (1973) (in Japanese); I Ojima, *J. Syn. Org. Chem., Japan*, *32*, 687 (1974) (in Japanese); H. B. Kagan, *Pure Appl. Chem.*, *43*, 401 (1975). H. B. Kagan and J. C. Fiaud, *Topics in Stereochemistry*, E. L. Eliel and N. L. Allinger, Eds., Wiley Interscience, New York, Vol. 10, 175 (1978).
65. (a) T.-P. Dang and H. B. Kagan, *J. Am. Chem. Soc.*, *94*, 6429 (1972). (b) T. Hayashi, K. Yamamoto, and M. Kumada, *J. Organomet. Chem.*, *112*, 253 (1976). (c) T. Hayashi, K. Yamamoto, K. Kasuga, H. Omizu, and M. Kumada, *J. Organomet. Chem.*, *113*, 127 (1976). (d) I. Ojima, T. Kogure, M. Kumagai, S. Horiuchi, and T. Sato, *J. Organomet. Chem.*, *122*, 83 (1976).
66. J. D. Morrison, R. E. Burnett, A. M. Aguiar, C. J. Morrow, and C. Phillips, *J. Am. Chem. Soc.*, *93*, 1301 (1971).
67. T. Hayashi, T. Mise, S. Mitachi, K. Yamamoto, and M. Kumada, *Tetrahedron Lett.*, 1133 (1976); T. Hayashi, M. Tajika, K. Tamao, and M. Kumada, *J. Am. Chem. Soc.*, *98*, 3718 (1976).
68. B. D. Vineyard, W. S. Knowles, M. J. Sabacky, G. L. Bachman, and D. J. Weinkauff, *J. Am. Chem. Soc.*, *99*, 5946 (1977).
69. M. D. Fryzuk and B. Bosnich, *J. Am. Chem. Soc.*, *99*, 6262 (1977).
70. W. Beck and H. Menzel, *J. Organomet. Chem.*, *133*, 307 (1977).

71. M. Florini, G. M. Giongo, F. Marcati, and W. Marconi, *J. Mol. Catal.*, *1*, 451 (1975-1976).
72. K. Hanaki, K. Kashiwabara, and J. Fujita, *Chem. Lett.*, 489 (1978).
73. K. Achiwa, T. Kogure, and I Ojima, *Tetrahedron Lett.*, 4431 (1977); *Chem. Lett.*, 297 (1978); K. Achiwa, *Chem. Lett.*, 561 (1978).
74. J. Halpern, D. P. Riley, A. S. Chan, and J. J. Pluth, *J. Am. Chem. Soc.*, *99*, 8055 (1977).
75. P. Pino, G. Consiglio, C. Botteghi, and C. Salomon, in *Homogeneous Catalysis—II*, D. Forster and J. F. Roth, Eds., American Chemical Society *Advances in Chemistry* Series, Vol. 132, ACS, Washington, D.C., 1974; G. Consiglio and P. Pino, *Helv. Chim. Acta*, *52*, 642 (1976).
76. T. Hayashi, M. Tanaka, and I. Ogata, *Tetrahedron Lett.*, 295 (1977).
77. B. Bogdanovič, *Angew. Chem., Int. Ed.*, *12*, 954 (1973).
78. T. Aratani, Y. Yoneyoshi, and T. Nagase, *Tetrahedron Lett.*, 2599 (1977).

6

FURTHER DEVELOPMENTS

1. HETEROGENIZED HOMOGENEOUS CATALYSTS

The advantages of homogeneous and heterogeneous catalysis have been combined recently in investigations that heterogenize homogeneous catalysts. These systems have opened new areas for catalytic research. Features such as high selectivity, homogeneous active sites, and ease of chemical modification have been retained in these heterogenized homogeneous catalysts. The advantages of heterogeneous catalysis, such as ease in catalyst removal, recovery and reactivation, durability, and thermal or air stability, have also been retained.

Cross-linked polystyrene has been primarily used as the heterogenizing agent, which chemically binds with coordinating —PPh_2 or —C_5H_4 ("Cp") groups; suitable transition metals have then been complexed to these groups. Because these polymer reactions are heterogeneous, the chemical transformations are not complete, and generally, some part of the resin remains unmodified. The catalytic activity of a polystyrene-bound $RhCl(PPh_3)_2$ catalyst for hydrogenation, however, exceeds that of the free $RhCl(PPh_3)_3$ primarily because the steric crowding caused by the polymer chains tends to maintain coordinative unsaturation during catalysis.[1] Typical examples of the heterogenized homogeneous catalysis (or polymer-bound or anchored catalysts) are shown in Table 20.

Table 20. Some Examples of Polymer-Bound Metal Complexes

Cross-Linked Polystyrene as the Polymer

- ⁓⌬–P(Ph)(Ph)–Ni(CO)$_2$(PPh$_3$)
- ⁓⌬–CH$_2$–⌬–Ti(Cl)(Cl)(Cp)
- ⁓⌬–CH$_2$–P(Ph)(Ph)–RhCl(PPh$_3$)$_2$
- ⁓⌬–Cr(CO)$_3$
- ⁓⌬–P(Ph)(Ph)–RhH(CO)(PPh$_3$)$_2$
- ⁓(pyridyl)–Co(CO)$_n$
- ⁓[⌬–P(Ph)(Ph)]$_2$–Pd–(CH=CH with CO, O, CO)
- ⁓⌬–CH$_2$PPh$_2$Rh$_4$(CO)$_{11}$

Silica or Alumina as the Polymer

- (C$_3$H$_5$)$_2$Nb bound via two O to Si–O–Si surface
- (C$_3$H$_5$)$_2$Mo bound via two O to Al–O–Al surface
- R–Si(O)(CH$_2$CH$_2$PPh$_2$ → Co(CO)$_3$) with R, OH on adjacent Si–O–Si surface

174

A chiral diphosphine of the DIOP type has been also bound to polystyrene to give a heterogeneous enantioselective hydrogenation catalyst[2]:

$$RCH=C(CO_2H)(NHCOCH_3) \xrightarrow[\text{P-cat}]{H_2} R-CH_2-CH(CO_2H)(NHCOCH_3)$$

Ⓟ—cat = polystyrene-bound Rh complex with CH(O)(O)CH bridging CH$_2$PPh$_2$ groups coordinated to RhCl

	Optical Yield (%)
R = H	52–60
R = Ph	86

Regioselectivity in the hydroformylation reaction has been influenced by the use of a polymer-attached RhH(CO)(PPh$_3$)$_3$ catalyst, as shown in the example below.[3]

$$CH_2=CHCH_2CH_3 + H_2 + CO \longrightarrow CH_3(CH_2)_3CHO + CH_3CH_2CH(CHO)CH_3$$

Ⓟ—Rh(CO)(PPh$_3$)$_2$

Ratio of normal to branched isomer = 16.1 (at 120°)

Some catalytically active π-allyl-metal complexes [e.g., $(\eta\text{-allyl})_n M$, where M = Ti, Zr, Cr, or Mo] that have been bound to the surface Si—OH groups by a hydrolytic M—C cleavage reaction have been found to be very active in ethylene polymerization (50–160°) and in olefin metathesis (at 20–60°)[4]:

$$\equiv Si-OH + (\eta^3\text{-}C_3H_5)_4 Zr \longrightarrow (\equiv Si-O)_2 Zr(C_3H_5)_2 + C_3H_6$$

Amino groups on "Amino-Aerosil" ($[(NH_2CH_2CH_2)_2SiO]_n$) have been utilized to bind the acidic *tert*-phosphine Ph_2PCO_2H, which complexes with Rh(I), yielding a long-lived hydrogenation catalyst.[5]

Enzymes have been also chemically bound to polystryene to give "immobilized enzymes." Both hexokinase and glucose-6-phosphate isomerase have been bound to the same polymer to catalytically convert glucose first to glucose-1-phosphate and then to glucose-6-phosphate. Immobilization of enzymes has been extensively studied because the practical importance of these substances, and a wide variety of solid supports have been exploited [e.g., cellulose, starch, polyacrylamide, poly(amino acid), porous glass, and zeolites]. Techniques of immobilization (or polymer binding), including chemical covalent bond formation, ionic interaction (salt formation), and polymer coating (encapsulation), should be further developed to discover other useful catalysts.

Since various parts of these polymer-bound homogeneous catalysts have been readily modified to improve selectivity and activity, further progress in this field is expected. A study of the "polymer effect" in catalysts might give some insight into the mechanism of enzyme reactions.

2. ENZYME-LIKE CATALYSTS

As the function and structure of actual enzymes have become better understood, research on enzyme-mimicking artificial materials has begun to flourish. Simple micelles formed from synthetic surfactants have been found to accelerate some organic as well as inorganic reactions, and these micelles are now considered to function as enzyme active centers and also as biomembranes.[6]

When cationic or anionic surfactants are dissolved in aqueous or nonaqueous solution or in a mixture of these solvents, micelles will form, if the concentration of the surfactant exceeds the critical micelle concentration (CMC). Usually in aqueous micelle solutions the long aliphatic chains of the surfactant tend to attract each other by hydrophobic forces and form a spherical micelle (Fig. 55*a*). In contrast, in nonpolar solvents the polar ends (cationic or anionic) of the surfactant associate to form inverted (or reversed) micelles (Fig. 55*b*). When a somewhat polar reagent molecule is placed in such micelles, its geometry is affected by orientation with the high electric field at the micelle boundary. Therefore polar reactions are accelerated in the presence of a micelle. The reaction in micellar boundaries is similar to some enzymatic reactions because enzymes usually have a hydrophobic part as well as a very polar part to accelerate a reaction.

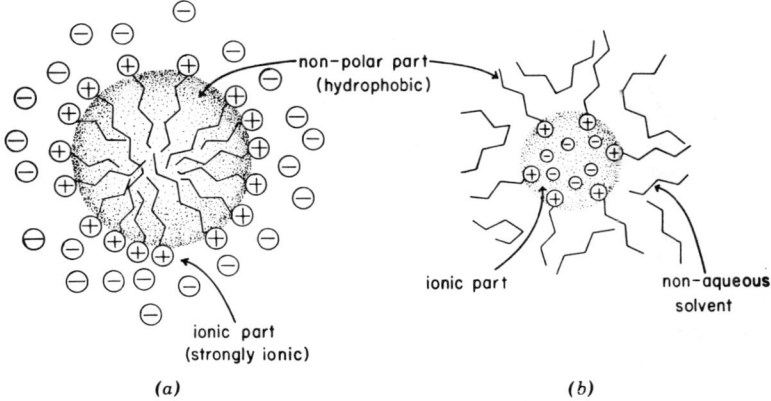

Fig. 55. (a) A spherical micelle in an aqueous medium. (b) An inverted micelle in a nonpolar medium.

$$\left[Co^I\left(\begin{array}{c}OH\\|\\N\diagdown\\C-CH_3\\\|\\C-CH_3\\N\diagup\\|\\O\end{array}\right)_2\right]^{\ominus} \xrightarrow[\text{surfactant}]{RX} \left[R-Co^{III}\left(\begin{array}{c}OH\\|\\N\diagdown\\C-CH_3\\\|\\C-CH_3\\N\diagup\\|\\O\end{array}\right)_2\right]$$

For example, alkylation of anionic cobaloxime ($[Co^I(dmg)_2]^-$) by alkyl halide is accelerated by cetyltrimethylammonium bromide 220-fold for EtBr and 420-fold for $Cl-CH_2CO_2^-$ compared to the uncatalyzed reaction.[7]

An inverse micelle formed in benzene causes a tremendous increase in the rate constant for some reactions. Aquation of an octahedral inert metal complex such as $[Cr(C_2O_4)]_3^{-3}$ is strongly catalyzed by alkylammonium carboxylate. The relative rate increase amounts to 5.4×10^6-fold.[8] Similarly, a ligand exchange reaction of vitamin B_{12} has been catalyzed by a mixture of dodecylammonium propionate and sodium bis(2-ethylhexyl)sulfosuccinate in benzene to yield a 10^2 to 10^4-fold increase (Figs. 56, 57; Scheme 35).

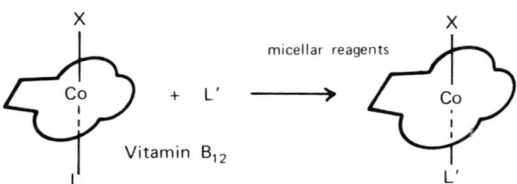

Scheme 35

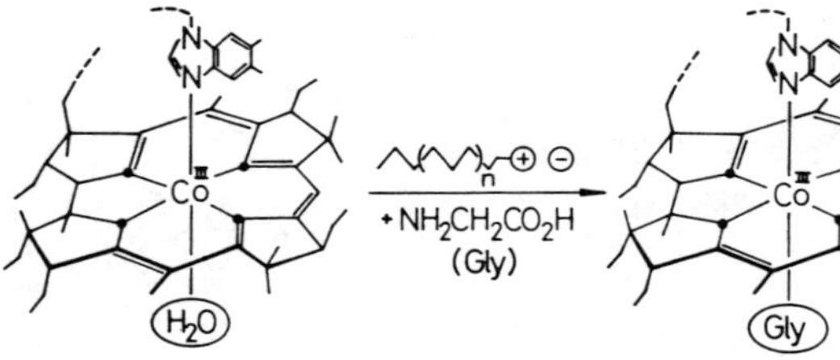

Fig. 56. A ligand exchange reaction at Co^{III} of vitamin B_{12a} [$BzmCo(H_2O)$] accelerated by micellar catalysis.

Crown ethers (e.g., synthetic cyclic polyethers, cycloamyloses, or cyclodextrins) have been thought to be similar to enzymes in having hydrophobic holes or pits of suitable sizes, which enhance the reaction of particular reactants. When the hydrophobic part of a reactant is just the size to fit in the hydrophobic hole of the crown ethers, the hydrophobic force[10] between the reactant and the crown ether holds the reactant in a particular steric arrangement. The functional groups at ideal positions relative to the reactant on the crown ether accelerate reactions. For example, α-cyclodextrin has been chemically modified by attaching organic functional groups

$$\left(e.g.,\ HN{\diagup\!\!\!\!\!-}N \right)$$

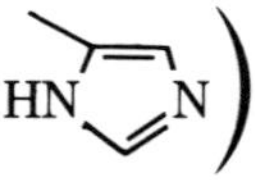

or a labile metal site. Organic acids having a hydrophobic hydrocarbon group that just fits in the dextrin can be hydrolyzed hundreds of times faster than in the absence of the crown ether. Table 21 lists the sizes of holes in the cycloamyloses.

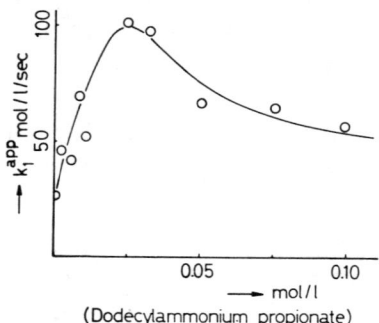

Fig. 57. The effect of surfactant concentrations on rates of micelle-catalyzed ligand exchange on aquated cobalamin.

Table 21. Hole Sizes of Typical Cycloamyloses (Å)

Example	Length	Width
cyclohexaamylose	6.7	4.5
cycloheptaamylose	~7	7.0
cyclooctaamylose	~7	8.5

An interesting polymer containing β-cyclodextrin (β-CD) as pendant groups has been prepared and found to promote the catalyzed hydrolysis of p-nitrophenylbenzoates by hydrophobic interactions as shown in Fig. 58.[11]

3. EXTREMELY SELECTIVE CATALYSTS

We have already seen that selectivity of catalysis has been enhanced by a suitable choice of active centers (cationic, anionic, or metallic) and surrounding environments (ligands, polypeptide chains, hydrophobic pockets, etc.). In an enzymatic hydrolysis by α-chymotrypsin, for example, one enantiomer of the ethyl ester of N-acetylphenylalanine is hydrolyzed at a very fast rate. This type of highly enantiomeric selectivity has not been observed with any artificially made catalysts. As the structures of enzyme active sites become better known, mechanisms of selective reactions of enzymes will be clarified in much more detail. Then the synthesis of effective active site models with the proper environment for a given reaction will be greatly accelerated. Accumulation of practical as well as theoretical knowledge of the chemistry of "complex" substances containing various organic functional groups and metal coordination spheres surely aids projected syntheses of such catalytically active species.

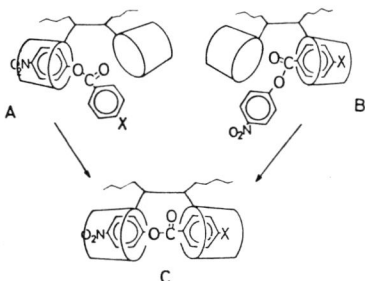

Fig. 58. The binding mode in the poly-β-cyclodextrin-p-nitrophenyl benzoate system. The crevices of the cyclodextrin rings accommodate the benzene ring of p-nitrophenylbenzoate.

$$\underset{\text{L-}(S)\text{-Isomer}}{\text{C}_6\text{H}_5\text{—CH}_2\text{—C}\underset{\text{NHCOCH}_3}{\overset{\text{H}}{\overset{|}{\underset{|}{\text{—CO}_2\text{C}_2\text{H}_5}}}}} \xrightarrow{\alpha\text{-chymotrypsin}} \text{C}_6\text{H}_5\text{—CH}_2\text{—C}\underset{\text{NHCOCH}_3}{\overset{\text{H}}{\overset{|}{\underset{|}{\text{—CO}_2\text{H}}}}}$$

$$\underset{\text{D-}(R)\text{-Isomer}}{\text{C}_6\text{H}_5\text{—CH}_2\text{—C}\underset{\text{NHCOCH}_3}{\overset{\text{CO}_2\text{C}_2\text{H}_5}{\overset{|}{\underset{|}{\text{—H}}}}}} \xrightarrow{\alpha\text{-chymotrypsin}} \mathbin{\!/\mkern-5mu/\!}$$

Tailoring the ligand to meet requirements of selectivity will be actively pursued. Extensive research utilizing chemically modified ligands has been done on nickel-complex-catalyzed linear dimerizations of propylene, mixed oligomerizations of monoolefins and conjugated dienes, and cyclooligomerizations of butadiene mainly at the Max Planck Institute (Mülheim).[12] These pioneering results indicate that suitable modification of ligands makes the selectivity to one particular product very high.

Optical resolution of chiral phosphine ligands, which at present are the most important among many ligands is now being actively investigated.[13] To enhance the selectivity, it seems better to have a chiral metal active center. For example, a four-coordinate tetrahedral metal complex with four different monodentate ligands has a chiral metal center. However the chiral center can be seriously disturbed during one catalytic cycle, resulting in racemization or disproportionation. Ligands with chiral coordinating atoms are the next choice. Some examples of this type have been found to be successful. Monodentate chiral phosphines do not always induce highly selective reactions because of ligand dissociation and free rotation around the metal-ligand bond. Chiral chelating ligands are much better in this respect. Already there have been a number of examples using these chiral chelates to induce high enantioselectivity. Monsanto's new chiral diphosphine chelate exemplifies this effectiveness (Chapter 5, Section 14).

Idealized examples of possible four-coordinate chiral metal complexes.

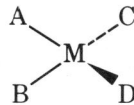

(a) Chiral metal center with four different ligands.

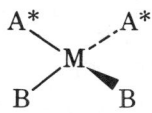

(b) Metal complexes with two chiral monodentate ligands.

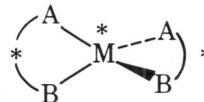

(c) Chiral metal center with two chelate ligands.

(d) Chiral metal center with two chiral chelate ligands.

4. ENERGY PROBLEMS

Homogeneous catalysis has great potential to help solve energy problems in the future. The photochemical production of hydrogen and oxygen from water has been reported to take place by way of a certain photocatalyst. This technique, although in the earliest stages of development, has given a clue to an important problem in effectively utilizing solar energy.

The direct photochemical reaction of water to produce two hydrogen atoms and one oxygen atom requires very high energy (220 kcal) or strong vacuum ultraviolet light (~130 nm). However only 58 kcal (or 500 nm) is necessary when gaseous hydrogen and oxygen molecules are the products. This photoenergy is near the energy maximum of the sunlight incident on earth. Water, however, is transparent in the visible region (340–800 nm) and does not absorb the solar energy. Photocatalysts having intense absorptions in the 500 nm region have been expected to be able to dissociate water into H_2 and O_2. However hydrogen and oxygen in the nascent state are very reactive, especially when both have been formed at the same time and place. Therefore even when the correct photoenergy has been absorbed by a photocatalyst, the simultaneous evolution of H_2 and O_2 has not been found to take place in a homogeneous solution.

Photodissociation of an H_2 molecule from transition metal dihydrides (H_2ML_n) has also been proposed as a means for obtaining hydrogen from water because some hydrides are formed easily on reaction with water, and some of them readily release hydrogen on irradiation. However generation of gaseous oxygen from water has not been successful because the oxidizing power of most

$$2[Co(CN)_5]^{3-} + H_2O \longrightarrow [HCo(CN)_5]^{3-} + [Co(CN)_5(H_2O)]^{2-} + OH^-$$

$$(H_2)_2Mo(dppe)_2 \xrightarrow[N_2]{} 2H_2 + (N_2)_2Mo(dppe)_2$$

known mononuclear transition metal complexes is weak. A binuclear dioxy-bridged complex

$$[(dipy)_2Mn\underset{O}{\overset{O}{\diamond}}Mn(dipy)_2]^{3+}$$

has decomposed on irradiation, giving gaseous oxygen, but it has not been catalytic when irradiated in water.[14]

The water-gas shift reaction

$$CO + H_2O \to CO_2 + H_2$$

has been widely achieved with heterogeneous catalysts at high temperatures. This process converts the energy content of carbon monoxide into hydrogen, which is a cleaner, more usable, and nontoxic form of energy. Recently the reaction has been found to proceed by way of a homogeneous catalyst at a relatively low temperature. Carbon monoxide has been reacted with water in the presence of $Ru_3(CO)_{12}$ in strongly alkaline solution to yield 5.6% H_2 and 6.5% CO_2 after 73 hr at 100°. Cluster complexes of complex structures seem to be promising for the "catalytic activation" of carbon monoxide.[15,16] A specially designed platinum(0) complex of $P(i-Pr)_3$ has been found to have high metal basicity and has been very active in the water-gas shift reaction at 100° in aqueous acetone.[17]

5. PHASE–TRANSFER CATALYSTS[18]

To promote ionic reactions with reactants that are present in two or more immiscible liquids (e.g., water and organic solvents), new types of homogeneous catalysts have been developed recently. These catalysts have been called phase-transfer catalysts, because they transfer one or more reactants from one phase to another, thereby facilitating a homogeneous reaction. Since this catalysis occurs by the partition of ions between two liquids, it has sometimes been called "ion-pair partition." The catalysts have been usually quaternary ammonium or quaternary phosphonium salts, which are soluble both in water and in organic solvents. An important difference from that of the micelle catalysis lies in the ability of phase-transfer catalysts to perform only by solubilization or modifying the ion partition by ion pairing rather than micelle formation.

Usually water-soluble reagents (mostly anionic reactants) have been reacted with one or more reactants in the organic phase with the aid of quaternary ammonium cations, which are partially soluble in the organic phase. Scheme 36 shows a water-soluble nucleophile A^- and an organic compound BX, which is soluble only in nonpolar organic solvents.

Scheme 36. Intended reaction, $A^- + BX \longrightarrow AB + X^-$, phase-transfer catalyst, P^+X^-.

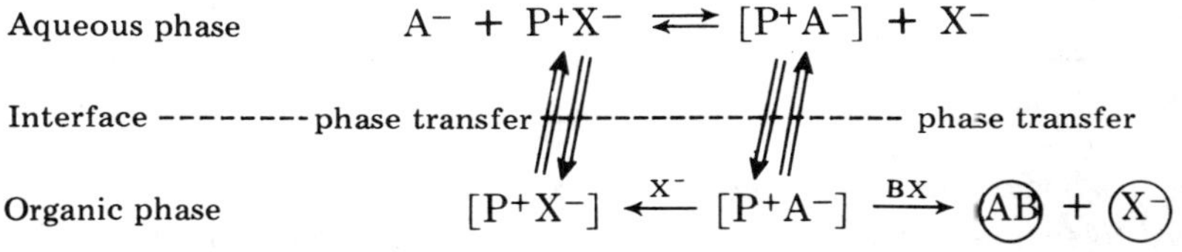

Aqueous phase $\quad A^- + P^+X^- \rightleftarrows [P^+A^-] + X^-$

Interface -------- phase transfer ———————————— phase transfer

Organic phase $\quad [P^+X^-] \xleftarrow{X^-} [P^+A^-] \xrightarrow{BX} \text{(AB)} + \text{(X}^-\text{)}$

The presence of ionic equilibria in the aqueous and organic phases and the solubility of the ionic aggregate $[P^+A^-]$ in the organic phase are the important points of this catalysis. In other words, the catalytic activity of phase-transfer catalysts depends on the high partition coefficient of the ion pair $[P^+A^-]$ between the aqueous and organic phases. The value of the coefficient must be larger than the value for the ion pair $[P^+X^-]$. Since the rate of the reaction depends on the partition coefficient of the ion pairs between the two phases, proper choice of solvent is important. Recommended solvents include CH_2Cl_2, $CHCl_3$, and o-$C_6H_4Cl_2$. In many cases faster rates have been observed when more polar but still water-immiscible solvents have been used. Nonpolar solvents are sometimes quite ineffective; for example, $[NEt_3(CH_2Ph)]^+$ has not been catalytically active in benzene for various S_N2 reactions.[19] Among the many ammonium or phosphonium cations, cations possessing larger side chains and more symmetric shapes have exhibited excellent catalytic activity (see Table 22). Among such excellent catalysts are $[N(n\text{-}Bu)_4]^+$, $[N(Me)(octyl)_3]^+$, $[NEt_3(n\text{-}C_{16}H_{33})]^+$, and $[P(n\text{-}Bu)_4]^+$. Considering prices and cation stability, $[N(Me)(octyl)_3]^+Cl^-$ has been the catalyst of choice.

The activity order of anionic nucleophiles (A^-) generally parallels their "hardness" (i.e., F > OH > Cl > Br > I). Using RX in alkylations with a phase-transfer catalyst, the order is RCl > RBr > RI, which is contrary to that usually observed without the phase-transfer catalyst.

Kinetic data show that reactions induced by phase-transfer catalysts proceed more rapidly than do the corresponding uncatalyzed "classical" reactions con-

Table 22. The Influence of Chain Length on Catalyst Activity in the Elimination Reaction $PhCH_2CH_2Br \rightarrow PhCH=CH_2$ [a]:

Catalyst $[CH_3-(CH_2)_n NEt_3]^+Br^-$	Yield of Styrene at 60° (%)
$n = 1$	3
2	7.2
3	12
4	50
5	53
7	44
11	38

[a] See ref. 18b.

ducted in a homogeneous medium. Thus the addition of a suitable phase-transfer catalyst to an aqueous/organic two-phase system usually accelerates the rate on the order of 10^4 to 10^9. Phase-transfer catalysis also allows some reactions in anhydrous solvents to be conducted in the presence of a large amount of water, which is another important advantage. The generation of dichlorocarbene from $CHCl_3$ and aqueous KOH is a remarkable example of this property.

A number of examples of the phase-transfer catalysis have been reported, and some are listed below according to reaction type[18]:

1. Oxidation: $RCH=CH_2 + KMnO_4 \xrightarrow{\text{in } C_6H_6} RCO_2H$

2. Reduction: $R-CHBr-CHBr-R \xrightarrow{Na/Na_2S_2O_3} RH=CHR$

3. Addition: $Me_3S^+I^- + RCHO \rightarrow R-\underset{O}{\triangle} + Me_2S$
   20–90%

4. Elimination: $PhCH_2CH_2Br \xrightarrow{NaOH} PhCH=CH_2$
   100%

5. Substitution: naphthol-OH $+ PhCH_2Cl \xrightarrow{OH^-}$ naphthyl-OCH_2Ph
   90%

6. Esterification: $RCO_2Na + CH_2Cl_2 \rightarrow (RCO_2)_2CH_2$
   60–90%

7. Carbene formation: cyclohexene + $CHCl_3 \xrightarrow{KOH}$ norcarane-7,7-dichloride 98%

8. Wittig reaction:

$$R_2C=O + CH_2(R'')-P(OEt)_2(=O) \xrightarrow{NaOH} R_2C=CHR''$$

55–80%

These techniques should become increasingly important in the future, and phase-transfer catalysis may become a standard technique in organic synthesis.[20]

SELECTED READINGS

A) Cyclodextrins

D. W. Griffith and M. L. Bender, "Cycloamyloses as Catalysts," *Adv. Catal.*, 23, 209 (1973).

B) Micellar Catalysts

J. H. Fendler and E. J. Fendler, *Catalysis in Micellar and Macromolecular Systems*, Academic Press, New York, 1975.

C) Heterogenized Homogeneous Catalysts

A. L. Robinson, "Homogeneous Catalysis: Anchored Metal Complexes," *Science*, 194, 1261 1976.

D) Phase-Transfer Catalysts

W. P. Weber and G. W. Gokel, *Phase Transfer Catalysis in Organic Synthesis*, Springer, Berlin, 1977.

R. Alan Jones, "Applications of Phase Transfer Catalysis in Organic Synthesis," *Aldrichimica Acta*, 9, 35, 1976.

REFERENCES

1. Reviews: J. C. Bailar, *Catal. Rev., 10,* 17 (1974); A. M. Michalska and D. E. Webster, *Platinum Metal Rev., 18,* 65 (1974); C. U. Pittman, Jr., and L. R. Smith, in *Organo Transition Metal Chemistry,* M. Tsutsui and Y. Ishii, Eds., Plenum Press, New York, 1975, pp. 143; F. R. Hartley and P. N. Vezey, *Adv. Organomet. Chem., 15,* 189 (1977); C. U. Pittman, Jr., and G. O. Evans, *Chem. Tech., 1973,* 560; C. U. Pittman, Jr., *Organomet. React. Synt. 6,* 1 (1977).
2. N. Takaishi, H. Imai, C. A. Bertelo, and J. K. Stille, *J. Am. Chem. Soc., 100,* 264 (1978).
3. C. U. Pittman, Jr., and R. M. Hanes, *J. Am. Chem. Soc., 98,* 5402 (1976).
4. J. P. Candlin and H. Thomas, in *Homogeneous Catalysis*—II, D. Forster and J. F. Roth, Eds., American Chemical Society *Advances in Chemistry* Series, Vol. 132, ACS, Washington, D.C., 1974, p. 212.
5. L. Horner and F. Schumacher, *Ann.,* 633 (1976).
6. J. H. Fendler and E. J. Fendler, *Catalysis in Micellar and Macromolecular Systems,* Academic Press, New York, 1975; J. H. Fendler, *Acc. Chem. Res., 9,* 153 (1976); E. H. Cordes, *Reaction Kinetics in Micelles,* Plenum Press, New York, 1973; E. H. Cordes and C. Gitler, *Prog. Bioorg. Chem., 2,* 1 (1973).
7. R. J. Allen and C. A. Bunton, *Bioinorg. Chem., 5,* 241 (1976).
8. C. J. O'Connor, E. J. Fendler, and J. H. Fendler, *J. Am. Chem. Soc., 95,* 600 (1973).
9. J. H. Fendler, F. Nome, and H. C. Van Woert, *J. Am. Chem. Soc., 96,* 6745 (1974).
10. W. P. Jencks, *Catalysis in Chemistry and Enzymology,* McGraw-Hill, New York, 1969, p. 393.
11. A. Harada, M. Furue, and S. Nozakura, *Macromolecules, 9,* 705 (1976).
12. P. W. Jolly and G. Wilke, *The Organic Chemistry of Nickel,* Vol. II, Academic Press, New York, 1975.
13. K. Tani, L. D. Brown, J. Ahmed, J. A. Ibers, M. Yokota, A. Nakamura, and S. Otsuka, *J. Am. Chem. Soc., 99,* 7876 (1977).
14. M. Calvin, *Science, 184,* 375 (1974).
15. R. M. Laine, R. G. Rinker, and P. C. Ford, *J. Am. Chem. Soc., 99,* 252, (1977).
16. C.-H. Cheng, D. E. Henricksen, and R. Eisenberg, *J. Am. Chem. Soc., 99,* 252, (1977); *Chem. Week,* April 19, 1978, p. 63.
17. S. Otsuka and T. Yoshida, *J. Am. Chem. Soc., 99,* 2134 (1977); T. Yoshida, Y. Ueda, and S. Otsuka, *J. Am. Chem. Soc., 100,* 3941 (1978).
18. (a) M. Makosza, A. Kacprowicz, and M. Fedorynski, *Tetrahedron Lett.,* 2119 (1975). (b) J. Dockx, *Synthesis,* 441 (1973). (c) E. V. Dehmlow, *Angew. Chem., Int. Ed., 13,* 170, (1974); *Chem. Tech.,* 210 (1975). (d) M. Makosza, *Pure Appl. Chem., 43,* 439, (1975). (e) G. W. Gokel and H. D. Durst, *Synthesis, 1976,* 168.
19. A. W. Herriott and D. Pieker, *J. Am. Chem. Soc., 97,* 2345 (1975).
20. S. Masamune, Ed., *Organic Syntheses,* Vol. 55, 1976, p. 91.

7

INDUSTRIAL APPLICATIONS

1. PETROCHEMICALS

Major industrial applications of homogeneous catalysis are listed in Table 23. Typical new developments in this area include (1) carbonylation of methanol by a soluble Rh(I) catalyst and (2) production of adiponitrile by a homogeneous nickel catalyst:

1. $CH_3OH \xrightarrow[\text{CO, 175°, 200 psi}]{[RhI_2(CO)_2]^-} CH_3CO_2H$

2. $CH_2=CH-CH=CH_2 + 2HCN \xrightarrow{Ni[P(OR)_3]_4/ZnCl_2}$
   $NC-CH_2CH_2CH_2CH_2CN$

The carbonylation of methanol[1] was developed by the Monsanto Company, and the process is now in operation. A very good yield of acetic acid (99% based on methanol, 90% based on CO) coupled with temperatures and pressures lower than the existing ones (e.g., the same reaction with CoI_2 catalyst operates at 210–250°, 7000–10,000 psi) has rendered this process superior in spite of the higher cost of the catalyst. When one considers the future importance of methanol as a chemical raw material, this method has an advantage that will make it one of the major industrial processes.

Table 23. Typical Examples of Industrially Important Homogeneous Catalyses

Reactants	Catalyst	Product	Condition
>C=C< /CO/H_2	$Co_2(CO)_8$ or [$Rh(CO)_2(PR_3)_2$]$_2$	>C—C< (H, CHO)	H_2/CO gas, 30–200 atm, 140–180° (CO), 100° (Rh)
>C=C< /CO/H_2O	$Ni(CO)_4$	>C—C< (H, CO_2H)	CO, 200 atm, in water, 270~320°
CH$_2$=CHCH=CH$_2$ (butadiene)	Ni(0)/P(O—Ph)$_3$	cyclooctadiene	30°
C_2H_4/AcOH/air	$Pd(OAc)_2$/CuCl	CH_2=CHOAc	100°
C_2H_4/H_2O/air	$PdCl_2$/CuCl/HCl	CH_3CHO	120–130°, 3 atm
Ph-CH$_3$/air	Co^{II} salt	Ph-CO_2H	110~120°, 2–3 atm
Ph-CO_2H/air	$Cu(O_2CPh)_2$	Ph-OH	190–250°
cyclohexane/air	$Co(OAc)_2$ or Co naphthenate	cyclohexanone + cyclohexanol	125~165°, 8~15 atm

Homogeneous Catalysis, B. J. Luberoff, Ed., American Chemical Society *Advances in Chemistry* Series, Vol. 70, ACS, Washington, D.C., 1968; *Homogeneous Catalysis—II*, J. F. Roth and D. Forster, Eds., ACS *Advances in Chemistry* Series, Vol. 132, 1974.

A recent DuPont process for the production of adiponitrile from butadiene and hydrogen cyanide has illustrated the potential of homogeneous catalysis:

$$CH_2=CH-CH=CH_2 \xrightarrow{HCN} CH_3-CH=CH-CH_2-CN \xrightarrow{isomerization} CH_2=CH-CH_2-CH_2-CN$$

$$NC-CH_2-CH_2-CH_2-CH_2-CN \xleftarrow{HCN}$$

Adiponitrile
↓
Hexamethylenediamine

Zerovalent nickel phosphite complexes have been found to be excellent homogeneous catalysts for the isomerization process and also for the subsequent second hydrocyanation process.[3] Using a nickel(0) complex with an organic

Petrochemicals

nitrile and a particularly bulky phosphite as the ligand, terminal hydrocyanation proceeds smoothly:

$$\text{CH}_2=\text{CHCH}_2\text{CH}_2\text{CN} + \text{HCN} \xrightarrow{6°} \text{adiponitrile,}$$

Catalyst system:

$$[(\text{NCCH}_2\text{CH}_2\text{CH}_2\text{CH}_2\text{CN})\text{Ni}[P(-O-C_6H_4-CH_3)_3]_3/P(-O-C_6H_4-CH_3)_3/\text{ZnCl}_2]$$

Protonation of the zerovalent nickel atom to give a cationic hydridonickel complex is followed by an olefin insertion, which seems to be the crucial point for the terminal cyanation because if the protonation occurs at the olefin, only a branched product (internal cyanation) can result. The steric effect of a bulky phosphite ligand also seems to prefer the terminal rather than the internal cyanation. With the successful attainment of the desired selectivity in this reaction, this hydrocyanation process seems to be economically superior to other existing processes.[4]

Dicobalt octacarbonyl also has catalyzed the hydrocyanation of olefins at 130°, but the resulting doubly hydrocyanated product from butadiene is not adiponitrile but a branched dinitrile[2]:

$$\text{CH}_2=\text{CHCH}=\text{CH}_2 + 2\text{HCN} \xrightarrow[130°, \ 40 \ \text{hr}]{\text{Co}_2(\text{CO})_8} \text{branched dinitrile with two CN groups}$$

Improvement of existing homogeneous catalysts is presently an active field of research. For example, synthesis of ethanol from methanol by conventional "oxo" conditions has become interesting because methanol and synthesis gas (CO and H_2) remain relatively inexpensive raw materials:

$$\text{CH}_3\text{OH} \xrightarrow[\text{CoI}_2 \ \text{or} \ \text{Co(OAc)}_2]{\text{CO/H}_2} \text{C}_2\text{H}_5\text{OH} + \text{CH}_3\text{CHO}$$

Air oxidation of hydrocarbons utilizing soluble transition metal salts will continue to be an important synthetic route for organic acids. Adipic acid has been produced in *one step* from cyclohexane with Co(OAc)_2 as the catalyst[5]:

$$\text{C}_6\text{H}_{12} + \text{O}_2 \xrightarrow{\text{Co(OAc)}_2} \text{HO}_2\text{C}-\text{CH}_2\text{CH}_2\text{CH}_2\text{CH}_2-\text{CO}_2\text{H}$$

Air oxidation of butane to acetic acid has also been improved so greatly that it can be used on an industrial scale[6]:

$$CH_3-CH_2-CH_2-CH_3 \xrightarrow{O_2} 2\,CH_3CO_2H$$

The regioselectivity of hydroformylation ("oxo" reaction) has been significantly enhanced by the use of special *tert*-phosphines as ligands or rhodium-phosphine complexes at catalysts.[7-9] Generally the yield of the economically more important linear aldehyde has been increased.

$$R-CH=CH_2 \rightarrow R-CH_2-CH_2CHO + R-CH(CHO)-CH_3$$

The high cost of rhodium catalysts has been more than compensated by the advantage of operating at lower pressures ($\sim$30 atm) and at lower temperatures (80–100°). The use of certain bulky *tert*-phosphine ligands increases the selectivity of the straight-chain product. The effect of a polymer support on selectivity and activity has not been completely clarified, but catalyst separation has been eased and selectivity for straight-chain products has been increased by using polystyrene-supported *tert*-phosphine or polyvinylpyridine ligands with $Co_2(CO)_8$ in hydroformylation reactions.[10]

2. COAL

The conversion of coal into CO is now a very important process, and selective hydrogenation of CO is being investigated actively. Metal clusters are novel homogeneous catalysts for various reactions.[11] For example, the reduction of carbon monoxide to methane[12], methanol[13], ethylene glycol[13], and others has been reported recently.

$$CO + H_2 \xrightarrow{Ir_4(CO)_{12}/AlCl_3} CH_4$$

$$\xrightarrow[220°, 1000\text{ atm}]{[Rh_6(CO)_{15}H]^-} \begin{array}{c} CH_2-CH_2 \\ | \quad | \\ OH \quad OH \end{array}$$

The stoichiometric formation of methanol from CO and H_2 has also been achieved[14a,b]:

$$(C_5Me_5)_2Zr(CO)_2 + H_2 \xrightarrow{HCl} CH_3OH + CO + (C_5Me_5)_2ZrCl_2$$

The mechanism of hydrogenation of CO has been studied by preparing intermediate metal complexes containing formyl group $(-CHO)$[14c,d] or $\eta^2\text{-}CH_2O$ ligand.[14e]

$$FeH(CO)_4^- + P(OPh)_3 \longrightarrow \left\{[(PhO)_3P](OC)_3Fe-C\begin{array}{c} \diagup O \\ \diagdown H \end{array}\right\}^-$$

$$Re_2(CO)_{10} + LiBHEt_3 \longrightarrow [(OC)_9Re_2CHO]^-$$

$$Os(CO)_3(PR_3)_2 \xrightarrow{CH_2O} \begin{array}{c} OC \underset{|}{\overset{PR_3}{\diagdown}} O \\ Os \\ OC \underset{PR_3}{\diagup} CH_2 \end{array} \longrightarrow \begin{array}{c} OC \underset{|}{\overset{PR_3}{\diagdown}} H \\ Os \\ OC \underset{PR_3}{\diagup} CHO \end{array}$$

The activation of a coordinated CO molecule for further reactions readily occurs in early transition metal complexes. For example, η^2-coordination of acyl group $(-CO-R)$ has recently been found.[14f,g]

Homogeneous catalysts show potential for solving environmental problems in the future. Catalytic conversion of poisonous NO and CO gases into non-poisonous N_2O and CO_2 has been successfully performed using $[RhCl_2(CO)_2]^-$ as a homogeneous catalyst in an aqueous acidic ethanol solution[15]:

$$2NO + CO \rightarrow N_2O + CO_2$$

3. FINE CHEMICALS

The high selectivity of homogeneous catalyst is frequently utilized in the fine chemicals industry, where delicate control of organic reactions is important. For example, asymmetric hydrogenation is currently being used for the production of L-dopa; a drug used to treat Parkinson's disease.[16] This pioneering undertaking may be followed by many other highly selective homogeneous catalysts. The enantioselective hydrogenation technique will undoubtedly be extended to reactions that give optically active natural products of various kinds, or their precursors.

$$\text{CH}_3\text{O-}\underset{\text{HO}}{\bigcirc}\text{-CH=C(NHCOCH}_3\text{)-CO}_2\text{H} \xrightarrow[\text{(2) Hydrolysis}]{\text{enantioselective catalyst} \atop \text{(1) H}_2} \text{CH}_3\text{O-}\underset{\text{HO}}{\bigcirc}\text{-CH}_2\text{-}\overset{*}{\text{CH}}(\text{NH}_2)\text{-CO}_2\text{H}$$

L-Dopa

Production of chrysanthemic acid, an essential component of a natural insecticides, has been made possible by Cu complexes with specially designed chiral ligands. The cyclopropanation has yielded the production in 90% enantiomeric excess when R = menthyl (see Chapter 5).[17]

$$\text{(alkene)} + \text{N}_2\text{CHCO}_2\text{R} \xrightarrow{\text{chiral Cu catalyst}} \text{(cyclopropane-CO}_2\text{R)} + \text{N}_2$$

Selective synthesis of naturally occurring useful terpenes using some transition metals combined with specially designed ligands has been actively examined. For example, regioselective catalytic dimerizations or trimerizations of isoprene have given a particular isomer of the terpene or a precursor of the terpene[18]:

$$\text{isoprene} \xrightarrow[\text{PhONa/AcOH}]{\text{PdCl}_2(\text{PR}_3)_2} \longrightarrow \longrightarrow \text{Geraniol}$$

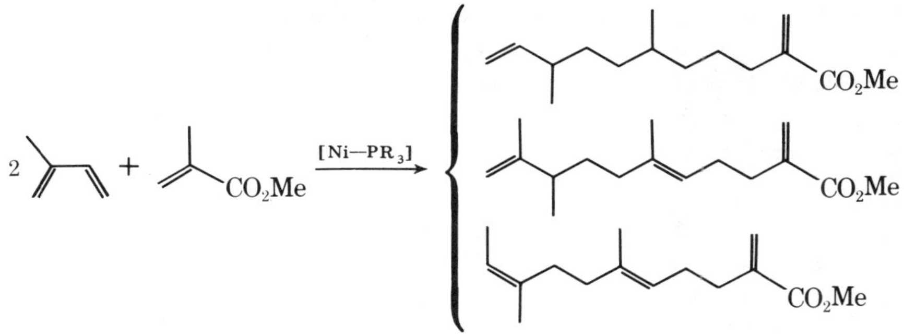

Fine Chemicals

Scheme 37

Co-oligomerizations using conjugated dimers, acetylenes, functionalized olefins, and organic carbonyl compounds have given a wide variety of products that are valuable as organic intermediates. Scheme 37 shows an example.[19] Exploitation of homogeneous catalysts for a particular process should become increasingly important.[20]

Homogeneous organic catalysts probably will be used for accelerating known organic reactions because faster reactions are more economical. There should also be further improvement of immobilized enzymes, and these new approaches may find extensive use in the selective production of natural products.

Homogeneous catalysts have been frequently utilized to achieve organic synthetic reactions involving organometallic reagents of magnesium, aluminum, lithium, and zinc. The simplest case in this category is the cross-coupling reaction between Grignard reagents and aryl or alkenyl halides catalyzed by phosphine-nickel catalysts [e.g., $NiCl_2(Ph_3P)_2$] developed by Tamao and Kumada[21]:

$$RMgX + arylX \rightarrow R-Ar$$

Similar cross-coupling reactions between RZnCl and alkenyl halides have been utilized to construct natural products as follows[22]:

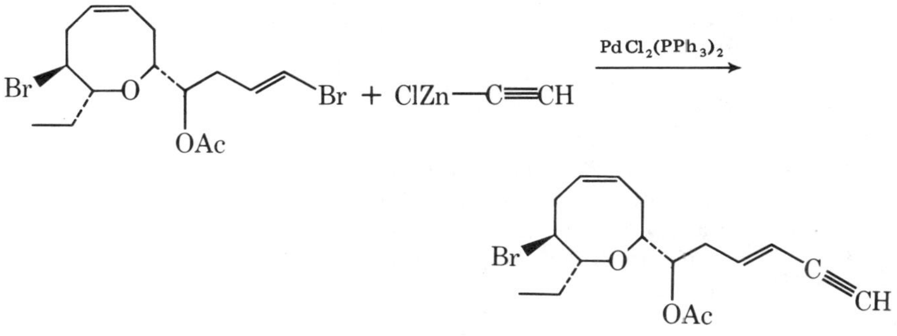

Homogeneous catalyses by Cp_2Zr species have been utilized to prepare trisubstituted olefins with high regio-, stereo-, and chemoselectivities as is shown in Scheme 38. Recent interest in the production of physiologically active terpenoid compounds may, therefore, provide the impetus for the commercialization of these selective catalytic reactions.[23]

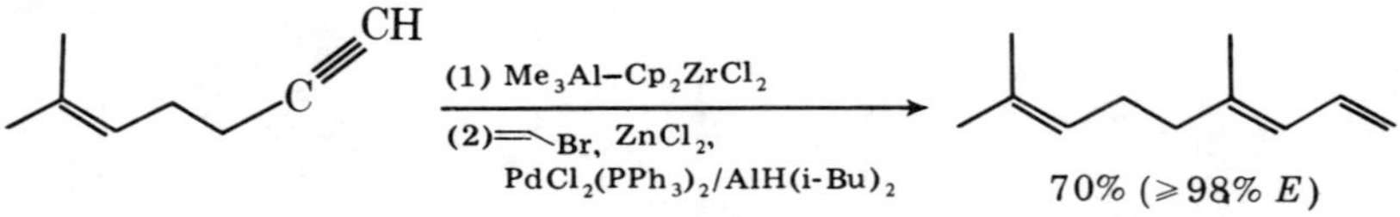

$$\text{olefin} \xrightarrow[\substack{(2) =\!\!\!\!\diagdown_{Br},\ ZnCl_2, \\ PdCl_2(PPh_3)_2/AlH(i\text{-}Bu)_2}]{(1)\ Me_3Al\text{-}Cp_2ZrCl_2} \text{product} \quad 70\%\ (\geqslant 98\%\ E)$$

Scheme 38

Micelles, cyclopolyethers, cryptates, and similar macromolecular catalysts have been employed to accelerate many organic reactions, especially in the synthesis of fine chemicals. Now the utility of these homogeneous catalysts is just beginning to be understood. Enzymes have been also used advantageously for the same purposes in some important lock-and-key reactions. Critical evaluation of these subjects is beyond the scope of this book.

REFERENCES

1. (a) J. F. Roth, *Chem. Tech.*, 600 (1971). (b) J. P. Grove, *Hydrocarbon Processing*, November 1976, p. 76. (c) F. E. Paulek and J. F. Roth, *Chem. Commun.*, 1578 (1968); *Chem. Week*, August 6, 1975, p. 23.
2. P. Arthur, Jr., D. C. England, B. C. Pratt, and G. M. Whitman, *J. Am. Chem. Soc.*, 76, 5364 (1954).
3. (a) W. C. Drinkard, Jr. (DuPont), U.S. Patent 3,496,218, *Chem. Abstr.*, 69, 26810 (1968). (b) C. A. Tolman, C. M. King, and W. C. Seidel, German Patent 2,237,703, *Chem. Abstr.*, 78, 135,700 (1973). (c) B. W. Taylor and H. E. Swift, *J. Catal.*, 26, 254 (1972).
4. *Mosaic*, September–October 1976, p. 28.
5. K. Tanaka, *Hydrocarbon Processing*, November 1974, p. 114.
6. R. P. Lowry and A. Aguilo, *Hydrocarbon Processing*, November 1974, p. 110.
7. R. Lai and E. Ucciani, in *Homogeneous Catalysis—II*, D. Forster and J. F. Roth, Eds., American Chemical Society *Advances in Chemistry* Series, Vol. 132, ACS, Washington, D.C., 1974, p. 1.
8. R. Kummer, H. J. Nienburg, H. Hohenschutz, and M. Strohmeyer, in *Homogeneous Catalysis—II*, D. Forster and J. F. Roth, Eds., American Chemical Society *Advances in Chemistry* Series, Vol. 132, ACS, Washington, D.C., 1974, p. 19.
9. R. Fowler, H. Connor, and R. A. Baehl, *Chem. Tech.*, 772 (1976).
10. H. B. Gray, *Chem. Eng. News*, March 1, 1976, p. 14.
11. (a) M. G. Thomas, E. L. Muetterties, R. O. Day, and V. W. Day, *J. Am. Chem. Soc.*, 98, 4645 (1976). (b) E. L. Muetterties, *Bull. Soc. Chim. Belg.*, 84, 959 (1975); 85, 451 (1976); *Science*, 196, 938 (1977).

References

12. (a) E. L. Muetterties, M. G. Thomas, and B. F. Beier, *J. Am. Chem. Soc.*, 98, 1296 (1976). (b) G. Henrici-Olivé and S. Olivé, *Agnew, Chem., Int. Ed.*, 15, 136 (1976). (c) M. A. Vannice, *Catal. Rev.*, 14, 153 (1976). (d) G. C. Demitras and E. L. Muetterties, *J. Am. Chem. Soc.*, 99, 2796 (1977).
13. (a) W. E. Walker, J. B. Cropley, and R. L. Pruett, *Chem. Tech.*, January 2, 1974; Netherlands Patents 7,407,383 and 7,407,412. (b) L. Kaplan, U.S. Patent 3,944,588. (c) J. W. Rathke and H. M. Feder, *J. Am. Chem. Soc.*, 100, 3623 (1978). (d) R. L. Pruett, *Ann. N.Y. Acad. Sci.*, 295, 239 (1977).
14. (a) J. M. Manriquez, D. R. McAlister, R. D. Sanner, and J. E. Bercaw, *J. Am. Chem. Soc.*, 98, 6733 (1976). (b) J. M. Manriquez, D. R. McAlister, R. D. Sanner, and J. E. Bercaw, *100*, 2716 (1978). (c) C. P. Casey and S. M. Neumann, *J. Am. Chem. Soc.*, 100, 2544 (1978). (d) J. A. Gladysz and W. Tamm, *J. Am. Chem. Soc.*, 100, 2545 (1978). (e) K. L. Brown, G. R. Clark, C. E. L. Headford, K. Marsden, and W. R. Roper, *J. Am. Chem. Soc.*, 101, 503 (1979). (f) J. M. Manriquez, P. J. Fagan, T. J. Marks, C. S. Day, and V. W. Day, *J. Am. Chem. Soc.*, 100, 7113 (1978). (g) F. Calderazzo, *Angew. Chem., Int. Ed.*, 16, 229 (1977).
15. (a) R. Eisenberg and C. D. Meyer, *Acc. Chem. Res.*, 8, 26 (1975). (b) D. E. Henricksen and R. Eisenberg, *J. Am. Chem. Soc.*, 98, 4662 (1976).
16. W. S. Knowles, M. J. Sabacky, and B. D. Vineyard, in *Homogeneous Catalysis—II*, D. Forster and J. F. Roth, Eds., American Chemical Society *Advances in Chemistry* Series, Vol. 132, ACS, Washington, D.C., 1974, p. 294. W. C. Christ et al., *J. Am. Chem. Soc.*, 101, 4406 (1979).
17. T. Aratani, Y., Yoneyoshi, and T. Nagase, *Tetrahedron Lett.*, 2599 (1977).
18. M. Hidai, H. Ishiwatari, H. Yagi, E. Tanaka, K. Onozawa, and Y. Uchida, *Chem. Comm.*, 170 (1975); J. P. Neilan, R. M. Lain, N. Gortese, and R. F. Heck, *J. Org. Chem.*, 41, 3455 (1976).
19. E. Klein, F. Thömel, H. Struse, P. Heimbach, and H. Schenkluhn, *Ann.*, 352 (1976); K.-J. Plöner and P. Heimbach, *Ann.*, 54 (1976).
20. H. Bandmann, P. Heimbach, and A. Roloff, *J. Chem. Res. (S)*, 261 (1977).
21. (a) K. Tamao, K. Sumitani, and M. Kumada, *J. Am. Chem. Soc.*, 94, 4374 (1972). (b) K. Tamao, Y. Kiso, K. Sumitani, and M. Kumada, *J. Am. Chem. Soc.*, 94, 9268 (1972). (c) M. Kumada, in *Organo-transition Metal Chemistry*, M. Tsutsui and Y. Ishii, Eds., Plenum Press, New York, 1975, p. 211.
22. E. Negishi, A. O. King, and N. Okukado, *Chem. Commun.*, 683 (1977).
23. (a) D. E. Van Horn and E. Negishi, *J. Am. Chem. Soc.*, 100, 2252 (1978). (b) E. Negishi, J. Okukado, A. O. King, D. E. Van Horn, and B. L. Spiegel, *J. Am. Chem. Soc., 100*, 2254 (1978).

Index

ABX pattern, 103
Acetaldehyde, 102
Acetamidocinnamic acid, 162
Acetoxylation, 146
(η^2-Acetylene) Zn(II), 98
Achiral, 30
Acid, catalysis, 110
   catalyzed process, 110
Acid-base, catalysis, 39
   interactions, 4
Activated, 67
Activation parameters, 22
Acylmetal carbonyls, 94
Addition, 1, 2, 75
   product, 45
Adiponitrile, 187, 188
Aggregate, 183
Alkoxy(alkyl)carbene, 94
Alkyl, aluminum, 141
   ammonium carboxylate, 177
   diazoacetates, 27
   isocyanides, 149
   rhodium, 123
   sodium, 140
η^3-Allyl, complexes, 101
   (η^3-Cyclopentadienyl) nickel, 133
   metal, 145
   metal complex, 143
Amino-aerosil, 176
Ammonia, 96
Antarafacial reaction, 82
Antibonding, 101
(η^6-Arene)chromium complexes, 97
Aromatic nucleus, 46
(η^1-Aryl)chromium compounds, 97
Aspartic acid, 114

Associative, mechanisms, 114
   reactions, 114
Asymmetric, centers, 29
   synthesis, 128
Atactic polypropylene, 141
ATP, 115
Autoxidation, 147

Back-donation, 63, 102
Base-catalyzed hydrolysis, 111
η^6-Benzene, 35
Betaine, 122
Bis-η^3-allyl-type, 135
d-Block transition metal, 36
Boltzmann's constant, 22
Boron trifluoride, 33
t-Butyl radical, 57

d-Camphor, 165
Carbenes, 80, 95
   dimers, 84
   mechanism, 152
Carbenic insertions, 79
Carbenoid reactions, 150
Carbonic anhydrase, 118
$tert$-Carbonium, 97
Carbonium cation, 139
Carbonylation, 77, 130, 187
Carbyne, 83
Catalyst life, 10
Catalytic, activation, 182
   sites, 10
Catechol, 148
Cationic polymerization, 139
C_{2v} distortion, 42
Cetyltrimethylammonium bromide, 177

$CH(AsF_5)_{0.064}$, 144
$CH_3{}^{13}COMn(CO)_5$, 80
Chelating diphosphines, 74, 96
Chelation, 74
Chemisorption, 9
Chemoselectivities, 194
Chiral center, 180
   chelating ligands, 180
   diphosphines, 162
   σ carbon, 81
   tert-phosphine, 165
Chrysanthemic acid, 167
α-Chymotrypsin, 179
Classical reactions, 183
Clusters, 55, 82, 182
$Co(acac)_3 PPh_3/AlEt_2 (OEt)$, 156
Coal, 190
Cobaloxime, 25, 177
Cobinamide, 50
Co, $CoCl(Ph_3P)$, 145. *See also* $HCo(CO)_4$
   $Co_2(CO)_8$, 129, 190
   $CoH(N_2)(PPh_3)_3$, 156
CO elimination, 80
Coenzyme B_{12}, 98
Coenzymes, 117
Complex reactions, 16
Condensation, 109
   product, 45
Cone angles, 62
Conjugated, dienes, 180
   polyene, 82
Co-oligomerizations, 193
Coordinated, ligands, 92
   dissociation, 60
   effect, 95
   equilibria, 64
   number, 37
η^2-Coordinated O_2, 149
η^2 Coordination, 102
Coordinatively unsaturated, 131, 142
$Co^{III}(R'CO_2)_2$, 148
Cossee mechanism, 142
Cp_2Zr, 194
Critical micelle concentration, 176

Crossed-mutual, 38
Crown ethers, 178
Cryptates, 194
Crystalline polymer, 145
Cu-containing oxidase, 122
Cumene, 147
Curtin-Hammett principle, 32
Cyclic nonpolar transition state, 81
Cycloaddition, 38, 77, 81
   amyloses, 178
   dextrins, 4, 178
   α-cyclodextrin, 178
   β-cyclodextrin, 179
   dimerization, 82
   octatetraene, 137
   oligomerization, 87, 137, 180
   polyethers, 194
   propanation, 27, 150, 165
   reversion, 82, 86
   tetramerization, 138
(η^5-Cyclopentadienyl)tricarbonylmolybdenum(II), 98
Cytochromes, 120
   c^{III}, 55

DA interactions, 39
Decarboxylation, 117
De-insertion, 75
Diacetylperoxide, 58
Dialkylcarbene, 84
   magnesium, 140
   cis-dialkyl(dipyridyl)nickel, 73
Diastereomers, 29
Diastereoselection, 30
Diastereoselectivity, 26
Diazenyl, 156
Diazoalkanes, 84
Diazo compounds, 117
Diborane, 76
Dichlorocarbene, 184
Diels-Alder reaction, 81
1,3-Dienes-d_2, 125
1,4-Dienes, 125
Diethyl(dipyridyl)nickel, 72
Diffusion-controlled, 53

Index

Dihydrides, 181
Dihydridoplatinum, 74
α-Diketones, 149
Dimerization, 57
Dimetallocyclobutane, 84
Dimethyl, 1,2-dideuterosuccinate, 77
  N,N-dimethylformamide(DMF), 150
  fumarate, 77
Diphenyl, bis(triphenylphosphine)nickel, 72
  carbene, 84
  (l-neomenthyl)phosphine, 161
Dipositive metal, 122
Disilacyclobutanes, 84
Disproportionation, 151
Dissociable protons, 46
Dissociative dimerization, 60
cis-Divinylcyclobutane, 87
Dodecylammonium propionate, 177
Donor-acceptor, 38
  interactions, 4
L-dopa, 192
Doping, 144
$d\pi$-$p\pi$, 84
Dynamic equilibrium, 100

Edge positions, 162
Edward's equation, 51
Effective oxidation state, 36
Electrocyclic reactions, 81
Electron, acceptors, 144
  microscope, 142
  d, 36
  spin resonance(esr)spectroscopy, 10
  transfer catalysis, 55
  transfer redox reactions, 119
  transporting, 120
  withdrawing substituents, 63, 140
Electrophilic, bimolecular substitution, 114
  interactions, 45
Electropositive metal, 63
Elementary, processes, 45
  reactions, 60, 85
Elimination, β, 78, 124
  geminal, 72

reductive, 72
Enantiofaces, 27
  selection, 167
Enantiomer, 179
Enantiomeric excess, 32
Enantioselective, 27
  catalytic reactions, 159
  selectivity, 13, 26, 126
Enantiotopic, 28
Encapsulation, 176
End-on, 95
Energetic, excitation, 54
  stabilization, 53
Energy barrier, 78
Enzymatic oxygenation, 95
Enzyme-like catalysts, 176
  mimicking, 176
Epoxidation, 150
Epoxides, 95
Equibinary polymer, 143
Equilibrium constants, 52
Excited dioxygen, 60
Extended x-ray absorption fine structure (EXAFS), 159
Extremely selective catalysts, 179

Face positions, 162
Faraday's constant, 120
$Fe(CO)_3$, 90
  $Fe(CO)_3 (\eta^4\text{-}C_8H_8)$, 89
Ferredoxins, 120
Ferrocenyl ketone, 30
Fine chemicals, 192
First, equilibrium constant, 61
  order reactions, 15
Fischer-Tropsch synthesis, 150
Five-coordinate, 37
Flavin adenine dinucleotide, 120
Fluxionality, 88
Forbidden, 77
Formal oxidation, 36
  state, 65
Formyl group, 191
η^1-Formylmethyl, 103
Four-coordinate, 37

Franck-Condon principle, 54
Free radicals, 56
Freevalence state, 142
Frequency factor, 22
Friedel-Crafts reaction, 46
Frontier orbitals, 34

Gas liquid chromatography, 32
gem-dihaloalkanes, 84
Geometry, 92
Glucose, 176
   1-phosphate, 176
   6-phosphate, 176
   6-phosphate isomerase, 176
Glutamic acid, 113
Group transfer reactions, 122

H-abstraction reaction, 29
Hard, 102
   nucleophiles, 49
   substrates, 49
Hardening, 102
$HCo(CO)_4$, 130
Hein's bis(η^6-arene)chromium complex, 97
Heterogeneity, 24
Heterogeneous catalysis, 9
Heterogenized homogeneous catalysts, 9, 173
Hexokinase, 176
HO_2^-, 149
HOMO, 39
   LUMO, 38
L-homocysteine, 122
Homogeneous catalysis, 9
   heterogeneous catalysts, 9
Hydration, 116
Hydrazine, 96, 156
β-Hydride, abstraction, 100
   elimination, 77
Hydrido, *cis*-hydridoalkyl complex, 127
   complexes, 76
   metal cation, 68
   olefin-complex, 124
   rhodium, 132

Hydroboration, 76
Hydroformylation, 130, 163, 190
Hydrogenations, 125
   tetrahapto(η^4), 126
   trihapto(η^3), 126
   1,4, 129
Hydrogen peroxide, 95
Hydrolysilylation, 128
Hydrolysis, 109
Hydrolytic enzymes, 112
Hydroperoxides, 150
Hydrophobic, 176, 178
Hydroxy, acids, 111
   imide, 94
   η^1-β-Hydroxyethylpalladium, 103

Immiscible liquids, 182
Immobilization, 176
Immobilized enzymes, 10, 176
Infrared(ir) spectroscopy, 10
Inhomogeneity, 10
Insertion, 75
   CO, 71, 80
Interactions, 45
   acid-base, 41, 46
   concerted, 112
   electrophilic, 47
   orbital, 46
   radical, 56
   σ, 102
Intramolecular catalysis, 111
   reaction, 97
Ionic elimination, 115
Ion-pair partition, 182
$IrH_5(PR_3)_2$, 91
Iron-heme enzymes, 122
Irreversible reductive elimination, 132
Irving-Williams series, 47
Isomerization, *cis-trans*, 60
   geometric, 60
Isonitriles, 95
Isostructural, 75
Isotactic, 141
Isotope effect, 102
Isotopic, labels, 12

Index

substitution, 88

β-Ketoacids, 117
Kinetics, 10
  control, 28

L-Lactic dehydrogenase, 11
Lactonization, 111
Laser-excited Raman spectroscopy, 10
Lewis acid, 47, 117
Ligands, auxiliary, 49, 74, 84, 98, 101, 132
  chelating, 74
  dissociation, 60
  substitution, 47
Linear relationship, 15
Lock-and-key reactions, 194
Low-spin, 37
Low-valent, 64
  metal, 86
  metal complexes, 84
LUMO, 39
Lysozyme, 113

Macromolecular, 120
  catalysts, 194
l-Menthol, 32
Mercurinium ion, 116
Metal, alkyl complexes, 151
  carbene, 143, 150, 167
  carbene complexes, 84
  carbon bonds, 72
  clusters, 120
  electron density, 102
  ethylene bond, 63
  ligand, 54, 180
  ligand bonds, 41
Metallation, carbo-, 75
  hydro, 75
Metallocene, 92
  cycles, 86
  cycloheptatriene, 137
  cyclopentadiene, 138
  cyclopentadiene complexes, 137
  cyclopropane, 86

enzymes, 4, 109, 115
  porphyrins, 120
Metathesis, 150
L-Methionine, 122
(E)-β-Methylcinnamic acid, 161
S-Methylation, 123
Methyl, cobalamine, 123
  corrinoid, 123
  migration, 80
  2-methyl-1,4-pentadiene-1-d_2, 125
  N-Methyltetrahydrofolic acid, 123
  S-Methyltransferase, 122
M–H bonding, 78
Micelles, 4, 194
  boundary, 176
  inverse, 177
  spherical, 176
  systems, 120
Michaelis-Menten rate equation, 18
Mixed-oxidation-state complex, 55
MoD_2Cp_2, 77
MoH_3Cp_2, 77
MoO_3 (DMF), 150
Na_2MoO_4, 150
Molecular orbital (MO) theory, 46
Molybdenum-iron-sulfur cluster, 159
Mono, dentate, 180
  methyl muconate, 148
Monoolefins, 180
MO theory, 117

N-acetylphenylalanine, 179
Na_2MoO_4, 150
Nascent state, 181
Natural enzymes, 4
Na_2WO_4, 150
$(N_2)_2W(PMe_2Ph)_4$, 96
Negative catalysis, 56
NH_3, 155
$NiCl_2(PEt_3)_2/AlEt_3$, 125
Nicotine-adenine-dinucleotide, 120
Nitride carrier, 156
Nitrogenase, 159
Nitrogen, coordinated, 96
  fixation, 120, 154

molecular, 95
Nmr, 89, 101
   ^{13}C, 90
   ^{1}H, 90
   see also P Nmr spectroscopy
Nonmetallic enzymes, 4
Nonpolar solvents, 183
Nucleophilic, attack, 92
   interactions, 45, 110
Nucleophilicity, 111

O_2^-, 95
Octahedral, 37, 177
O_h structures, 37
O isotopic, 115
Olefin, η^2-olefin complexes, 102
   metathesis, 85, 175
   monoolefins, 180
   α, 148
   triolefin process, 151
Oligomerizations, 133, 180
   linear, 137
Oligomers, 133
Oligopeptides, 110
One-electron transfer, 59, 147
Optical, purity, 32
   resolution, 180
Optishift reagents, 32
Orbital, dπ, 69
   $\rho\pi$, 59
   symmetry, 33, 85, 86
Organic enzymes, 4
Organometallic, catalysts, 2
   methylcobalt, 123
   reagents, 193
Organosilicon compounds, 84
Organotransition metal complexes, 82
Orthopalladation, 146
Oxidases, 122
Oxidation-reduction, 147
Oxidative, addition, 61, 64, 131
   cleavage, 149
Oxidoreductases, 120
Oxo, conditions, 189

processes, 2
β-Oxoethyl, 103
Oxygenation, 59

P-450, 122
Panthothenic acid, 162
Paramagnetic, 59
Parkinson's disease, 192
P-*chiral tert*-phosphines, 161
Pd(OAc)$_2$, 146
Pericyclic reactions, 82
Peroxy, molybdenum, 150
   tungsten, 150
Phase-transfer catalysts, 182
Phenylacetylene, 138
Phosphine ligands, 180
   tert, 190
Photo, catalyst, 181
   chemical, 37, 99
   chemical production, 181
   chemical reactions, 15
   dissociation, 181
π, acceptors, 35
   acidic, 92
   acids, 35
   basic, 92
   donating, 35
   radicals, 59
Planck's constant, 22
p-nitrophenylbenzoates, 179
P nmr spectroscopy, 162
   see also Nmr
Polar addition, 115
cis-polyacetylene, 114
Poly, methylene, 80
   pentenamer, 151
   peptide, 179
   styrene-bound, 173
   topal rearrangements, 88
   vinylene, 144
Polyaromatic compounds, 129
Polyhedral coordination spheres, 88
Polyhydrido complexes, 91
cis-Polyisoprene, 140
Polymer, anchored catalysts, 9

Index

coating, 176
effect, 176
supported homogeneous catalysts, 9
Polymerization, 77, 139
 ring-opening, 151
Porphyrin, 55
pπ-dπ, 101
Predissociation, 68
Preequilibrium, 68
Prismatic, 88
Pro, R, 28
 S, 28
Prochiral, 27
Propagating species, 139
Propagation, 140
Propylene polymerization, 141
Protolytic, cleavage, 116
 reductive cleavage, 96
Protonation, 45
Pseudo-first-order, 48
 reactions, 16
Pyridoxal phosphate, 122
Pyrolysis, 86

Quaternary phosphonium, 182
o-Quinones, 149

Racemization, 122, 180
Radicals, anion, 58
 cation, 58, 148
 chain, 71
 chain-transfer-compounds, 140
 coupling, 57
 density, 58
 heterocoupling, 57
 homocoupling, 57
 initiated reaction, 140
 like species, 56
 mechanism, 129
 metal-centered, 59
 stabilized, 58
Rate-determining steps, 10, 20, 48
Reaction, 2+2, 85
 4+2, 85
 energetics, 10

Recombination, 57
Redox, properties, 2
 reaction, 12
Reductive elimination, 54
Regioselective, 192
 insertion, 100
Regioselectivity, 26, 175
Reppe's vinylation reaction, 92
Retention, 81
Reversible σ-π rearrangement, 100
RhCl(PPh$_3$)$_3$, 126
Rigidity, 88
[Ru(NH$_3$)$_5$N$_2$]$^{2+}$, 156

Salicylaldehyde, 122
Schiff base, 118, 159
Second-order reaction, 16
Selectivity, 11, 25
Self-associative reaction, 84
Semiconductors, 144
Side-chain chiral phosphines, 161
Side-on, 149
σ, accepting, 35
 alkenylmercuric complex, 116
 π rearrangements, 97
 symmetry, 101
 type donation, 63
 type donor, 33
Silaethylenes, 84
Singlet-triplet, 60
Six-coordinate, 37
S_N2, mechanism, 70
 reactions, 183
Sodium, bis(2-ethylhexyl) sulfosuccinate, 177
 naphthalenide, 96
Soft, 102
 and hard, 49
 nucleophiles, 49
 substrates, 49
Softening, 102
sp^2 carbon, 34
Specificity, 11
Spin-paired, 36
Stereochemistry, 33

Stereochemically nonrigid, 88
Stereoregular polymerization, 140
Stereoselective, 25
Stereospecific, 25
Stoichiometric reactions, 10
Super, acids, 2
  electrophiles, 2
  nucleophiles, 51
  nucleophilicity, 123
Surfactant, 176
Symmetric, homogeneous catalysts, 4
  shapes, 183
  spherically, 37
  totally, 37
Symmetry, forbidden, 78
  relaxation, 125
  rules, 87
Syndiotactic polypropylene, 141
Synthetic organic catalysts, 4

Terminal olefins, 124, 133, 143
Terpene, 192
Tetracyanoethylene, 35
Tetrahedral, 37
Thermal, distortion, 41
  isomerization, 16
  reaction, 80
Thermally forbidden, 86
α-$TiCl_3$, 142
Transamination, 122
f-Transition metals, lanthanides, actinides, 36
Tricoordinated carbon, 34
Triisopropoxides, 156
Thioglycerol, 155
Three-center transition state, 81
Trans effect, 74
Transpolymer, 144
$trans$-R_2Ni[P(cyclohex)$_3$]$_2$[51], 73
Triad elements, 74

Trimetallobicyclobutane, 83
Triolefin process, 151
Triphenyltris(tetrahydrofuran) chromium
  (III), 97
Tryptophan pyrolase, 148
Two-phase system, 145

$U(C_8H_8)_2$, 36

Vanadium(II)/catechol/water, 155
Vaska's, complex, 65
  iridium complex, 35
VCl_4/$AlEt_2Cl$, 143
Vinyl, alcohol, 145
  η^2-vinylalcohol, 103
  monomers, 56
η^2-Vinylsilylether, 103
Vitamin B_{12s}, 50
Vol'pin-Shur system, 155

Wacker process, 2, 145
Water, gas shift reaction, 182
  immiscible solvents, 183
  soluble reagents, 183
Wilkinson's, catalyst, 20, 25
  complex, 159
$W(N_2)_2(PMe_2Ph)_4$, 156
  Na_2WO_4, 150
  $(N_2)_2W(PMe_2Ph)_4$, 96
Woodward-Hoffman rules, 77, 82

X-ray diffraction analysis, 90

Zeroth-order reactions, 14
Zerovalent nickel, 87
  phosphite complexes, 188
Ziegler catalysis, 2, 141
  Natta catalysts, 79
  processes, 2

RAYMOND H. FOGLER LIBRARY

DATE DUE

BOOKS ARE SUBJECT TO
RECALL AFTER TWO WEEKS

MAY 14 1982